Meat Hygiene and Technology

Meat Hygiene and Technology

O.P. Thakur

AMIGA PRESS. INC
Delhi -110051

AMIGA PRESS. INC
H.O :- 28-A, Indra Park, Chander Nagar
Delhi :- 110051 (India)
Tel :- 011-22504028, +91-98736-57000
Email :- amigapress@gmail.com

This edition is published by Amiga Press.Inc

ISBN : 978-93-84533-09-0

First Published 2018

Printed and bound in India by Thomson Press Ltd., New Delhi

Preface

Meat processing technology comprises the steps and procedures in the manufacture of processed meat products. Processed meat products, which include various different types and local/regional variations, are food of animal origin, which contribute valuable animal proteins to human diets.

Meat processing hygiene is part of Quality Management (QM) of meat plants and refers to the hygienic measures to be taken during the various processing steps in the manufacture of meat products. Regulatory authorities usually provide the compulsory national framework for food/meat hygiene programmes through laws and regulations and monitor the implementation of such laws. At the meat industry level, it is the primary responsibility of individual enterprises to develop and apply efficient meat hygiene programmes specifically adapted to their relevant range of production. Operations in meat processing plants comprise the manufacture of value-added meat products from primary products of meat origin and non-meat origin. There are three principles of meat hygiene, which are crucial for meat processing operations.

Meat is mainly composed of water and protein, and is usually eaten together with other food. It is edible raw, but is normally eaten after it has been cooked and seasoned or processed in a variety of ways. Unprocessed meat will spoil within hours or days. Spoilage is caused by the practically unavoidable infection and subsequent decomposition of meat by bacteria and fungi, which are borne by the animal itself, by the people handling the meat, and by their implements.

Microbial meat spoilage or food poisoning through meat can be prevented if the microbial load/bacterial contamination, which occurs during slaughtering and meat handling, is kept as low as possible. The key for achieving this is strict meat hygiene including an uninterrupted cold chain throughout the entire meat production and handling chain.

In meat-product manufacturing, the basic processing technologies, such as cutting and mixing, are accompanied by various additional treatments and procedures, depending on the type and quality of the final product. Such treatments involve curing, seasoning, smoking, filling into casings or rigid containers, vacuum packaging, cooking or canning/sterilization. Due to the importance of these procedures, suitable and up-to-date techniques for carrying out these processes and the equipment needed are described in separate chapters but are also referred to in the manual in connection with the respective product groups.

Meat processing technologies were developed particularly in Europe and Asia. The European technologies obviously were more successful, as they were disseminated and adopted to a considerable extent in other regions of the world - by way of their main creations of burger patties, frankfurter-type sausages and cooked ham. The traditional Asian products, many of them of the fermented type, are still popular in their countries of origin. But Western-style products have gained the upper hand and achieved a higher market share than those traditional products. In Asia and Africa, there are a number of countries where meat is very popular but the majority of consumers reject processed meat products.

This book is broadly covering fresh meat and aquatic foods, their processing, preservation, packaging standards and biotechnological applications in this specialized field.

—*Editor*

Contents

1

Meat: An Introduction

Meat is animal flesh that is eaten as food. Humans are omnivorous, and have hunted and killed animals for meat since prehistoric times. The advent of civilization allowed the domestication of animals such as chickens, sheep, pigs and cattle, and eventually their use in meat production on an industrial scale.

Meat is mainly composed of water and protein, and is usually eaten together with other food. It is edible raw, but is normally eaten after it has been cooked and seasoned or processed in a variety of ways. Unprocessed meat will spoil within hours or days. Spoilage is caused by the practically unavoidable infection and subsequent decomposition of meat by bacteria and fungi, which are borne by the animal itself, by the people handling the meat, and by their implements.

Meat consumption varies worldwide, depending on cultural or religious preferences, as well as economic conditions. Vegetarians choose not to eat meat because of ethical, economic, environmental, religious or health concerns that are associated with meat production and consumption.

Most often, *meat* refers to skeletal muscle and associated fat and other tissues, but it may also describe other edible tissues such as offal. Conversely, *meat* is sometimes used in a more restrictive sense – the flesh of mammalian species (pigs, cattle, lambs, etc.) raised and prepared for human consumption, to the exclusion of fish and other seafood, poultry or other animals.

Etymology

The word *meat* comes from the Old English word *mete*, which referred to food in general. The term is related to *mad* in Danish, *mat* in Swedish and Norwegian, and *matur* in Icelandic andFaroese, which also mean 'food'. The word *mete* also exists in Old Frisian (and to a lesser extent, modern West Frisian) to denote important food, differentiating it from *swiets* (sweets) and *dierfied* (animal feed).

History

Paleontological evidence suggests that meat constituted a substantial proportion of the diet of even the earliest humans. Early hunter-gatherers depended on the organized hunting of large animals such as bison and deer.

The domestication of animals, of which we have evidence dating back to the end of the last glacial period (c. 10,000 BC), allowed the systematic production of meat and the breedingof animals with a view to improving meat production.

A typical shoulder cut of lamb:

- Sheep, originating from western Asia, were domesticated with the help of dogs prior to the establishment of settled agriculture, likely as early as the 8th millennium BC. Several breeds of sheep were established in ancient Mesopotamia and Egypt by 3500–3000 BC. Presently, more than 200 sheep breeds exist.
- Cattle were domesticated in Mesopotamia after settled agriculture was established about 5000 BC, and several breeds were established by 2500 BC. Modern domesticated cattle fall into the groups*Bos taurus* (European cattle) and *Bos indicus* (zebu), both descended from the now-extinct aurochs. The breeding of beef cattle, cattle optimized for meat production as opposed to animals best suited for draught or dairy purposes, began in the middle of the 18th century.

A Hereford bull, a breed of cattle frequently used in beef production.

- Domestic pigs, which are descended from wild boars, are known to have existed about 2500 BC in modern-day Hungary and in Troy; earlier pottery from Jericho and Egypt depicts wild pigs. Porksausages and hams were of great commercial importance in Greco-Roman times. Pigs continue to be bred intensively as they are being optimized to produce meat best suited for specific meat products.

Other animals are or have been raised or hunted for their flesh. The type of meat consumed varies much between different cultures, changes over time, depending on factors such as tradition and the availability of the animals. The amount and kind of meat consumed also varies by income, both between countries and within a given country.

- Horses are commonly eaten in countries such as France Italy, Germany and Japan. Horses and other large mammals such as reindeer were hunted during the Late Paleolithic in western Europe.
- Dogs are consumed in China, South Korea and Vietnam. Dogs are also occasionally eaten in the Arctic regions. Historically, dog meat has been consumed in various part of the world, such asHawaii, Japan, Switzerland and Mexico.

- Cats are consumed in Southern China and Peru.
- Guinea pigs are raised for their flesh in the Andes.
- Whales and dolphins are hunted, partly for their flesh, in Japan, Alaska, Siberia, Canada, the Faroe Islands, Greenland, Iceland, Saint Vincent and the Grenadines and by two small communities in Indonesia.

Modern agriculture employs a number of techniques, such as progeny testing, to make animals evolve rapidly to acquire the qualities desired by meat producers. For instance, in the wake of well-publicised health concerns associated with saturated fats in the 1980s, the fat content of United Kingdom beef, pork and lamb fell from 20–26 percent to 4–8 percent within a few decades, due to both selective breeding for leanness and changed methods ofbutchery. Methods of genetic engineering aimed at improving the meat production qualities of animals are now also becoming available.

Even though it is a very old industry, meat production continues to be shaped strongly by the evolving demands of customers. The trend towards selling meat in pre-packaged cuts has increased the demand for larger breeds of cattle, which are better suited to producing such cuts. Even more animals not previously exploited for their meat are now being farmed, especially the more agile and mobile species, whose muscles tend to be developed better than those of cattle, sheep or pigs. Examples are the various antelope species, the zebra, water buffalo and camel, as well as non-mammals, such as the crocodile, emu and ostrich. Another important trend in contemporary meat production is organic farming which, while providing noorganoleptic benefit to meat so produced, meets an increasing demand for organic meat.

Growth and Development of Meat Animals

Agricultural science has identified several factors bearing on the growth and development of meat in animals.

Genetics

Trait	*Heritability*
Reproductive efficiency	2–10%
Meat quality	15–30%
Growth	20–40%
Muscle/fat ratio	40–60%

Several economically important traits in meat animals are heritable to some degree and can thus be selected for by animal breeding. In cattle, certain growth features are controlled by recessive genes which have not so far been controlled, complicating breeding. One such trait is dwarfism; another is the doppelender or "double muscling" condition, which causes muscle hypertrophy and thereby increases the animal's commercial value. Genetic analysis continues to reveal the genetic mechanisms that control numerous aspects of the endocrine system and, through it, meat growth and quality.

Genetic engineering techniques can shorten breeding programmes significantly because they allow for the identification and isolation of genes coding for desired traits, and for the reincorporation of these genes into the animal genome. To enable such manipulation, research is ongoing (as of 2006) to map the entire genome of sheep, cattle and pigs. Some research has already seen commercial application. For instance, a recombinant bacterium has been developed which improves the digestion of grass in the rumen of cattle, and some specific features of muscle fibres have been genetically altered.

Experimental reproductive cloning of commercially important meat animals such as sheep, pig or cattle has been successful. The multiple asexual reproduction of animals bearing desirable traits can thus be anticipated, although this is not yet practical on a commercial scale.

Environment

Heat regulation in livestock is of great economic significance, because mammals attempt to maintain a constant optimal body temperature. Low temperatures tend to prolong animal development and high temperatures tend to retard it. Depending on their size, body shape and insulation through tissue and fur, some animals have a relatively narrow zone of temperature tolerance and others (e.g. cattle) a broad one. Static magnetic fields, for reasons still unknown, also retard animal development.

Nutrition

The quality and quantity of usable meat depends on the animal's *plane of nutrition*, i.e., whether it is over- or underfed. Scientists disagree, however, about how exactly the plane of nutrition influences carcase composition.

The composition of the diet, especially the amount of protein provided, is also an important factor regulating animal growth. Ruminants, which may digest cellulose, are better adapted to poor-quality diets, but their ruminal microorganisms degrade high-quality protein if supplied in excess. Because producing high-quality protein animal feed is expensive, several techniques are employed or experimented with to ensure maximum utilization of protein. These include the treatment of feed with formalin to protect amino acids during their passage through the rumen, the recycling of manure by feeding it back to cattle mixed with feed concentrates, or the partial conversion of petroleum hydrocarbons to protein through microbial action.

In plant feed, environmental factors influence the availability of crucial nutrients or micronutrients, a lack or excess of which can cause a great many ailments. In Australia, for instance, where the soil contains limited phosphate, cattle are being fed additional phosphate to increase the efficiency of beef production. Also in Australia, cattle and sheep in certain areas were often found losing their appetite and dying in the midst of rich pasture; this was at length found to be a result of cobalt deficiency in the soil. Plant toxins are also a risk to grazing animals; for instance, sodium fluoracetate, found in some African and Australian plants, kills by disrupting the cellular metabolism. Certain man-made pollutants such as methylmercury and somepesticide residues present a particular hazard due to their tendency to bioaccumulate in meat, potentially poisoning consumers.

Human Intervention

Meat producers may seek to improve the fertility of female animals through the administration of gonadotrophic or ovulation-inducing hormones. In pig production, sow infertility is a common problem, possibly due to excessive fatness. No methods currently exist to augment the fertility of male animals. Artificial insemination is now routinely used to produce animals of the best possible genetic quality, and the efficiency of this method is improved through the administration of hormones that synchronize the ovulation cycles within groups of females.

Growth hormones, particularly anabolic agents such as steroids, are used in some countries to accelerate muscle growth in animals. This practice has given rise to the beef hormone controversy, an international trade dispute. It may also decrease the tenderness of meat, although research on this is inconclusive, and have other effects on the composition of the muscle flesh. Where castration is used to improve control over male animals, its side effects are also counteracted by the administration of hormones.

Sedatives may be administered to animals to counteract stress factors and increase weight gain. The feeding of antibiotics to certain animals has been shown to improve growth rates also. This practice is particularly prevalent in the USA, but has been banned in the EU, partly because it causesantibiotic resistance in pathogenic microorganisms.

Biochemical Composition

Numerous aspects of the biochemical composition of meat vary in complex ways depending on the species, breed, sex, age, plane of nutrition, training and exercise of the animal, as well as on the anatomical location of the musculature involved. Even between animals of the same litter and sex there are considerable differences in such parameters as the percentage of intramuscular fat.

Main Constituents

Adult mammalian muscle flesh consists of roughly 75 percent water, 19 percent protein, 2.5 percent intramuscular fat, 1.2 percent carbohydrates and 2.3 percent other soluble non-protein substances. These include nitrogenous compounds, such as amino acids, and inorganic substances such as minerals.

Muscle proteins are either soluble in water (sarcoplasmic proteins, about 11.5 percent of total muscle mass) or in concentrated salt solutions (myofibrillarproteins, about 5.5 percent of mass). There are several hundred sarcoplasmic proteins. Most of them – the glycolytic enzymes – are involved in the glycolytic pathway, i.e., the conversion of stored energy into muscle power. The two most abundant myofibrillar proteins, myosin and actin, are responsible for the muscle's overall structure. The remaining protein m6ass consists of connective tissue (collagen and elastin) as well as organelletissue.

Fat in meat can be either adipose tissue, used by the animal to store energy and consisting of "true fats" (esters of glycerol with fatty acids), or intramuscular fat, which contains considerable quantities of phospholipids and of unsaponifiable constituents such as cholesterol.

CRITICISMS AND RESPONSES OF MEAT

Morals

It has been argued that a moral community requires all participants to be able to make moral decisions, but animals are incapable of making moral choices (e.g., a tiger would not refrain from eating a human because it was morally wrong; it would decide whether to attack based on its survival needs, as dictated by hunger). Thus, some opponents of ethical vegetarianism argue that the analogy between killing animals and killing people is misleading. Humans are capable of culture, innovation and the sublimation of instinct in order to act in an ethical manner. Animals are not, and so are unequal to humans on a moral level. This does not excuse cruelty, but it implies animals are not morally equivalent to humans and do not possess the rights a human has. For example, killing a mouse is not the moral equivalent of committing homicide.

Benjamin Franklin describes his conversion to vegetarianism in chapter one of his autobiography, but then he describes why he (periodically) ceased vegetarianism in his later life:

...in my first voyage from Boston...our people set about catching cod, and hauled up a great many. Hitherto I had stuck to my resolution of not eating animal food... But I had formerly been a great lover of fish, and, when this came hot out of the frying-pan, it smelt admirably well. I balanc'd some time between principle and inclination, till I recollected that, when the fish were opened, I saw smaller fish taken out of their stomachs; then thought I, 'If you eat one another, I don't see why we mayn't eat you.' So I din'd upon cod very heartily, and continued to eat with other people, returning only now and then occasionally to a vegetable diet. So convenient a thing it is to be a reasonable creature, since it enables one to find or make a reason for everything one has a mind to do.

The response of ethical vegetarians is that humans have a choice, whereas animals do not. Hence if hurting or killing animals is not necessary for human survival or health (as it is necessary for obligate carnivores, such as felids), humans can decide to stop doing it, and to outlaw it.

The animal rights movement, sometimes called the animal liberation movement, animal personhood, or animal advocacy movement, is a social movement which seeks an end to the rigid moral and legal distinction drawn between human and non-human animals, an end to the status of animals as property, and an end to their use in the research, food, clothing, and entertainment industries.

Animal Welfare

Various programs operate in the U.S. that promote the notion that animals raised for food can be treated humanely. Spokesemen for the factory farmingindustry argue that the animals are better off in total confinement. For example, according to F J "Sonny" Faison, president of Carroll's Foods:

They're in state-of-the-art confinement facilities. The conditions that we keep these animals in are much more humane than when they were out in the field. Today they're in housing that is environmentally controlled in many respects. And the feed is right there for them all the time, and water, fresh water. They're looked after in some of the best conditions, because the healthier and [more] content that animal, the better it grows. So we're very interested in their well-being up to an extent.

In response, animal welfare advocates such as Farm Sanctuary argue that commodifying and slaughtering animals is incompatible with the definition of "humane". Animal ethicists such as Gary Francione have argued that reducing animal suffering is not enough; it needs to be made illegal andabolished.

Nonchordates

Peter Singer has pointed out that the ethical argument for vegetarianism may not apply to all non-vegetarian food. For example, any arguments against causing pain to animals would not apply to animals that do not have central nervous systems if they do not feel pain. It has also often been noted that it takes a lot more grain to feed a cow for human consumption than it takes to feed a human directly, but not all animals consume land plants (or other animals that consume land plants). Oysters, for example, consume underwater plankton and algae. In 2010, Christopher Cox wrote:

Biologically, oysters are not in the plant kingdom, but when it comes to ethical eating, they are almost indistinguishable from plants. Oyster farms account for 95 percent of all oyster consumption and have a minimal negative impact on their ecosystems; there are even nonprofit projects devoted to cultivating oysters as a way to improve water quality. Since so many oysters are farmed, there's little danger of overfishing. No forests are cleared for oysters, no fertilizer is needed, and no grain goes to waste to feed them—they have a diet of plankton, which is about as close to the bottom of the food chain as you can get. Oyster cultivation also avoids many of the negative side effects of plant agriculture: There are no bees needed to pollinate oysters, no pesticides required to kill off other insects, and for the most part, oyster farms operate without the collateral damage of accidentally killing other animals during harvesting. Cox went on to suggest that oysters would be acceptable to eat, even by strict ethical criteria, if they did not feel: "while you could give them the benefit of the doubt, you could also say that unless some new evidence of a capacity for pain emerges, the doubt is so slight that there is no good reason for avoiding eating sustainably produced oysters." Cox has added that, although he believes in some of the ethical reasons for vegetarianism, he is not strictly a vegan or even a vegetarian because he consumes oysters.

Animal and Plant Pain

Critics of ethical vegetarianism point out that there is no agreement on where to draw the line between organisms that can and cannot feel. Justin Leiber, a philosophy professor at Oxford University writes that:

Montaigne is ecumenical in this respect, claiming consciousness for spiders and ants, and even writing of our duties to trees and plants.Singer and Clarke agree in denying consciousness to sponges. Singer locates the distinction somewhere between the shrimp and the oyster. He, with rather considerable convenience for one who is thundering hard accusations at others, slides by the case of insects and spiders and bacteria, they pace Montaigne, apparently and rather conveniently do not feel pain. The intrepid Midgley, on the other hand, seems willing to speculate about the subjective experience of tapeworms ...Nagel ... appears to draw the line at flounders and wasps, though more recently he speaks of the inner life of cockroaches.

There are also some who argue that although only suffering animals feel anguish, plants, like all organisms, have evolved mechanisms for survival. No living organism can be described as "wanting" to die for another organism's sustenance. In an article written for the New York Times, Carol Kaesuk Yoon argues that:

When a plant is wounded, its body immediately kicks into protection mode. It releases a bouquet of volatile chemicals, which in some cases have been shown to induce neighboring plants to pre-emptively step up their own chemical defenses and in other cases to lure in predators of the beasts that may be causing the damage to the plants. Inside the plant, repair systems are engaged and defenses are mounted, the molecular details of which scientists are still working out, but which involve signaling molecules coursing through the body to rally the cellular troops, even the enlisting of the genome itself, which begins churning out defense-related proteins ... If you think about it, though, why would we expect any organism to lie down and die for our dinner? Organisms have evolved to do everything in their power to avoid being extinguished. How long would any lineage be likely to last if its members effectively didn't care if you killed them?

RED MEAT

Red meat in traditional culinary terminology is meat which is red when raw and not white when cooked. Red meat also includes the meat of most adultmammals.

Gastronomic

In gastronomy, red meat is darker-colored meat, as contrasted with white meat. The exact definition varies by time, place, and culture, but the meat from adult mammals such as cows, sheep, and horses is invariably considered red, while chicken and rabbit meat is invariably considered white. The meat of young mammals such as milk-fedveal calves, sheep, and pigs is traditionally considered white; while the meat of duck and goose is considered red. Game is sometimes put in a separate category altogether. Veal calves is traditionally a red meat if it is naturally fed (French: *viandes noires* — "black meats").

Nutritional

The old determinant of the nutritional definition of the color of meat is the concentration of myoglobin. The white meat of chicken has under 0.05%; pork and veal have 0.1–0.3%; young beef has 0.4–1.0%; and old beef has 1.5–2.0%.

According to the USDA all meats obtained from livestock (i.e., from mammals) are red meats because they contain more myoglobin than chicken or fish.

Nutrition

Red meat contains large amounts of iron, creatine, minerals such as zinc and phosphorus, and B-vitamins: (niacin, vitamin B_{12}, thiamin and riboflavin).Red meat is the richest source of lipoic acid, a powerful antioxidant.

Red meat contains small amounts of vitamin D. The liver contains much higher quantities than other parts of the animal.

Food Pyramid

The 1992 edition of the USDA food guide pyramid has been criticized for not distinguishing between red meat and other types of meat. The 2005 edition, MyPyramid, recommends lean forms of red meat.

Potential Health Risks

There is significant controversy about the health effects of consuming red meat. Part of this is due to the fact that studies (cited below) show that red meat is not a uniform product but rather seems to vary significantly in health effect based on fat content, processing and preparation. Many studies do not differentiate between these factors. More recent studies, such as a 2010 Harvard University study covering over one million people, have studied the effect of fresh (unprocessed) versus processed red meat and found processed meat to have significant health risks, but no effect for unprocessed red meat. A massive study conducted by the European Prospective Investigation into Cancer and Nutrition and published in 2013 which followed over 400,000 subjects showed increased mortality among subjects that consumed processed meat regularly, but no effect for unprocessed red meat. Yet another study linked consumption of processed meat to premature death (after adjusting for confounding factors), even though a consumption of a small amount of red meat appeared to have beneficial effects.

Colorectal Cancer

Due to the many studies that have found a link between red meat intake and colorectal cancer, the American Institute for Cancer Research and World Cancer Research Fund stated that there is convincing evidence that red meat intake increases the risk for colorectal cancer.

Professor Sheila Bingham of the Dunn Human Nutrition Unit attributes this to the haemoglobin and myoglobin molecules which are found in red meat. She suggests these molecules, when ingested trigger a process called nitrosation in the gut which leads to the formation of carcinogens. Others have suggested that it is due to the presence of carcinogenic compounds called heterocyclic amines, which are created in the cooking process.However, this may not be limited to red meat, since a study from the Harvard School of Public Health found that people who ate skinless chicken five times or more per week had a 52% higher risk of developing bladder cancer although not people who ate chicken with skin or other red meats such as hot dogs and Hamburgers.

A 2011 study of 17,000 individuals found that people consuming the most grilled and well-done meat had a 56 and 59% higher rate of cancer.

Other Cancers

There is suggestive evidence that red meat intake increases the risk of oesophageal, lung, pancreatic and endometrial cancer. As a result, WCRF recommends limiting intake of red meat to less than 300g (11 oz) cooked weight per week, "very little, if any of which to be processed."

Some studies have linked consumption of large amounts of red meat with breast cancer, stomach cancer, lymphoma, bladder cancer, lung cancer and prostate cancer (although other studies have found no relationship between red meat and prostate cancer).

A 2011 study of almost 500,000 participants found that those in the highest quintile of red meat consumption had a 19% increased risk of kidney cancer.

Cardiovascular Diseases

The postwar Seven Countries Study found a significant correlation between red meat consumption and risk of CHD and marked the beginning of our current understanding.

Many studies associate red meat consumption with cardiovascular diseases. Specifically red meat consumption is associated with ischemic heart disease, stroke, with greater intima-media thickness, (an indicator of atherosclerosis), acute coronary syndrome, A significant relationship between red meat and CHD has been found specifically for women,

Processed Meat Vs Unprocessed

The consensus on the role of red meat consumption to increased risk of cardiovascular diseases has changed in recent years. Studies that differentiate between processed and fresh red meat have failed to find a link between unprocessed red meat consumption and heart disease. A major Harvard University meta-study in 2010 involving over one million people who ate meat found that only processed meat had an adverse risk in relation to coronary heart disease. The study suggests that the "differences in salt and preservatives, rather than fats, might explain the higher risk of heart disease and diabetes seen with processed meats, but not with unprocessed red meats."

Mechanisms

Some mechanisms that have been suggested for why red meat consumption is a risk factor for cardiovascular disease include: its impact on serum cholesterol, that red meat contains arachidonic acid, heme iron, and homocysteine., its high content of saturated fat. Bacteria in the digestive tract of people who eat meat have been found to produce a spike in TMAO when supplied with carnitine (abundant in red meat). TMAO is a metabolite that promotes atherosclerosis, a thickening of the arteries.

Risk

A study suggests that eating 50g (less than 2oz) of processed meat per day increases risk of coronary heart disease by 42%, and diabetes by 19%. Equivalent levels of fat, including saturated fats, in unprocessed red meat (even when eating twice as much per day) did not show any deleterious effects, leading the researchers to suggest that "differences in salt and preservatives, rather than fats, might explain the higher risk of heart disease and diabetes seen with processed meats, but not with unprocessed red meats."

Those that eat more than 8 servings of red meat per month are 4.9 times more likely to have cardiac events than those eating less than four servings per month.

A 21-year follow up of about thirty thousand Seventh-day Adventists (adventists are known for presenting a "health message" that recommends vegetarianism) found that people who ate red meat daily were 60% more likely to die of heart disease than those who ate red meat less than once per week.

The risk of coronary disease due to high cholesterol can be mitigated by switching to a leaner red meat. According to one study, funded by the beef producers advocacy group, National Cattlemen's Beef Association, eating lean meat (bout red and white) produced nearly identical cholesterol, and triglyceride levels in both groups.

Controversial Studies

A 2009 study by the National Cancer Institute found a correlation between the consumption of red meat and increased mortality from cardiovascular diseases, as well as increased mortality from all causes. This study has been criticized for using an improperly validated food frequency questionnaire, which has been shown to have low levels of accuracy.

Diabetes

Red meat intake has been associated with an increased risk of type II diabetes. Interventions in which red meat is removed from the diet can lower albuminuria levels. Replacing red meat with a low protein or chicken diet can improve glomerular filtration rate.

Other findings have suggested that the association may be due to saturated fat, trans fat and dietary cholesterol, rather than red meat per se. An additional confound is that diets high in processed meat could increase the risk for developing Type 2 diabetes.

One study estimated that "substitutions of one serving of nuts, low-fat dairy, and whole grains per day for one serving of red meat per day were associated with a 16–35% lower risk of type 2 diabetes".

Obesity

The Diogenes project used data from ninety thousand men and women over about seven years and found that "higher intake of total protein, and protein from animal sources was associated with subsequent weight gain for both genders, strongest among women, and the association was mainly attributable to protein from red and processed meat and poultry rather than from fish and dairy sources. There was no overall association between intake of plant protein and subsequent changes in weight." They also found an association between red meat consumption and increased waist circumference.

A 1998 survey of about five thousand vegetarian and non-vegetarian people found that vegetarians had about 30% lower BMIs. A 2006 survey of fifty thousand women found that those with higher "western diet pattern" scores gained about two more kilograms over the course of four years than those who lowered their scores.

A ten-year follow up of 80,000 men and women found that "ten-year changes in body mass index was associated positively with meat consumption" as well as with weight gain at the waist. In a Mediterranean population of 8,000 men and women, meat consumption was significantly associated with weight gain. Data from the National Health and Nutrition Examination Survey showed "consistent positive associations between meat consumption and BMI, waist circumference, obesity and central obesity."

RED AND WHITE MEAT

Meat can be broadly classified as "red" or "white" depending on the concentration of myoglobin in muscle fibre. When myoglobin is exposed to oxygen, reddish oxymyoglobin develops, making myoglobin-rich meat appear red. The redness of meat depends on species, animal age, and fibre type: Red meat contains more narrow muscle fibres that tend to operate over long periods without rest, while white meat contains more broad fibres that tend to work in short fast bursts.

Generally, the meat of adult mammals such as cows, sheep, goats, and horses is considered red, while chickenand turkey breast meat is considered white.

Nutritional Information

All muscle tissue is very high in protein, containing all of the essential amino acids, and in most cases is a good source of zinc, vitamin B_{12}, selenium,phosphorus, niacin, vitamin B_6, choline, riboflavinand iron. Several forms of meat are high invitamin K_2, which is only otherwise known to be found in fermented foods , with nattohaving the highest concentration. Muscle tissue is very low in carbohydrates and does not contain dietary fiber. The fat content of meat can vary widely depending on the species and breed of animal, the way in which the animal was raised, including what it was fed, the anatomical part of the body, and the methods of butchering and cooking. Wild animals such as deer are typically leaner than farm animals, leading those concerned about fat content to choose game such as venison. Decades of breeding meat animals for fatness is being reversed by consumer demand for meat with less fat.

The table in this section compares the nutritional content of several types of meat. While each kind of meat has about the same content of protein and carbohydrates, there is a very wide range of fat content. It is the additional fat that contributes most to the calorie content of meat, and to concerns about dietary health.

Production

Meat is produced by killing an animal and cutting flesh out of it. These procedures are called slaughter andbutchery, respectively. There is ongoing research into producing meat *in vitro*, that is, outside of animals.

Attesting to the long history of meat consumption in human civilizations, ritual slaughter has become part of the practice of several religions. These rituals, as well as other pre-industrial meat production methods such as these used by indigenous peoples, are not detailed here. This section will instead provide an overview of contemporary industrialized meat production in dedicated slaughterhouses from cattle, sheep and pigs.

Transport

Upon reaching a predetermined age or weight, livestock are usually transported *en masse* to the slaughterhouse. Depending on its length and circumstances, this may exert stress and injuries on the animals, and some may die *en route*. Unnecessary stress in transport may adversely affect the quality of the meat. In particular, the muscles of stressed animals are low in water and glycogen, and their pH fails to attain acidic values, all of which results in poor meat quality. Consequently, and also due to campaigning by animal welfare groups, laws and industry practices in several countries tend to become more restrictive with respect to the duration and other circumstances of livestock transports.

Slaughter

Animals are usually slaughtered by being first stunned and then exsanguinated (bled out). Death results from the one or the other procedure, depending on the methods employed. Stunning can be effected through asphyxiating the animals with carbon dioxide, shooting them with a gun or a captive bolt pistol, or shocking them with electric current. In most forms of ritual slaughter, stunning is not allowed.

Draining as much blood as possible from the carcase is necessary because blood causes the meat to have an unappealing appearance and is a breeding ground for microorganisms. The exsanguination is accomplished by severing the carotid artery and the jugular vein in cattle and sheep, and the anterior vena cava in pigs.

Dressing and Cutting

After exsanguination, the carcass is dressed; that is, the head, feet, hide (except hogs and some veal), excess fat, viscera and offal are removed, leaving only bones and edible muscle. Cattle and pig carcases, but not those of sheep, are then split in half along the mid ventral axis, and the carcase is cut into wholesale pieces. The dressing and cutting sequence, long a province of manual labor, is progressively being fully automated.

Conditioning

Under hygienic conditions and without other treatment, meat can be stored at above its freezing point (–1.5 °C) for about six weeks without spoilage, during which time it undergoes an aging process that increases its tenderness and flavor.

During the first day after death, glycolysis continues until the accumulation of lactic acid causes the pH to reach about 5.5. The remaining glycogen, about 18 g per kg, is believed to increase the water-holding capacity and tenderness of the flesh when cooked. *Rigor mortis* sets in a few hours after death as ATPis used up, causing actin and myosin to combine into rigid actomyosin and lowering the meat's water-holding capacity, causing it to lose water ("weep"). In muscles that enter *rigor* in a contracted position, actin and myosin filaments overlap and cross-bond, resulting in meat that is tough on cooking – hence again the need to prevent pre-slaughter stress in the animal.

Over time, the muscle proteins denature in varying degree, with the exception of the collagen and elastin ofconnective tissue, and *rigor mortis* resolves. Because of these changes, the meat is tender and pliable when cooked just after death or after the resolution of *rigor*, but tough when cooked during *rigor*. As the muscle pigment myoglobin denatures, its iron oxidates, which may cause a brown discoloration near the surface of the meat. Ongoing proteolysis also contributes to conditioning. Hypoxanthine, a breakdown product of ATP, contributes to the meat's flavor and odor, as do other products of the discomposition of muscle fat and protein.

Additives

When meat is industrially processed in preparation of consumption, it may be enriched with additives to protect or modify its flavor or color, to improve its tenderness, juiciness or cohesiveness, or to aid with its preservation. Meat additives include the following:

- Salt is the most frequently used additive in meat processing. It imparts flavor but also inhibits microbial growth, extends the product's shelf life and helps emulsifying finely processed products, such as sausages. Ready-to-eat meat products normally contain about 1.5 to 2.5 percent salt.
- Nitrite is used in curing meat to stabilize the meat's color and flavor, and inhibits the growth of spore-forming microorganisms such as *C. botulinum*. The use of nitrite's precursor nitrate is now limited to a few products such as dry sausage, prosciutto or parma ham.
- Phosphates used in meat processing are normally alkaline polyphosphates such as sodium tripolyphosphate. They are used to increase the water-binding and emulsifying ability of meat proteins, but also limit lipid oxidation and flavor loss, and reduce microbial growth.
- Erythorbate or its equivalent ascorbic acid (vitamin C) is used to stabilize the color of cured meat.
- Sweeteners such as sugar or corn syrup impart a sweet flavor, bind water and assist surface browning during cooking in the Maillard reaction.
- Seasonings impart or modify flavor. They include spices or oleoresins extracted from them, herbs, vegetables and essential oils.
- Flavorings such as monosodium glutamate impart or strengthen a particular flavor.
- Tenderizers break down collagens to make the meat more palatable for consumption. They include proteolytic enzymes, acids, salt and phosphate.

- Dedicated antimicrobials include lactic, citric and acetic acid, sodium diacetate, acidified sodium chloride or calcium sulfate, cetylpyridinium chloride, activated lactoferrin, sodium or potassium lactate, or bacteriocins such as nisin.
- Antioxidants include a wide range of chemicals that limit lipid oxidation, which creates an undesirable "off flavor", in precooked meat products.
- Acidifiers, most often lactic or citric acid, can impart a tangy or tart flavor note, extend shelf-life, tenderize fresh meat or help with protein denaturationand moisture release in dried meat. They substitute for the process of natural fermentation that acidifies some meat products such as hard salami or prosciutto.

Spoilage and Preservation

The spoilage of meat occurs, if untreated, in a matter of hours or days and results in the meat becoming unappetizing, poisonous or infectious. Spoilage is caused by the practically unavoidable infection and subsequent decomposition of meat by bacteria and fungi, which are borne by the animal itself, by the people handling the meat, and by their implements. Meat can be kept edible for a much longer time – though not indefinitely – if proper hygiene is observed during production and processing, and if appropriate food safety, food preservation and food storage procedures are applied. Without the application of preservatives and stabilizers, the fats in meat may also begin to rapidly decompose after cooking or processing, leading to an objectionable taste known as warmed over flavor.

Methods of Preparation

Fresh meat can be cooked for immediate consumption, or be processed, that is, treated for longer-termpreservation and later consumption, possibly after further preparation. A common additive to processed meats, both for preservation and because it prevents discoloring, is sodium nitrite, which, however, is also a source of health concerns, because it may form carcinogenic nitrosamines when heated.

Meat is prepared in many ways, as steaks, in stews, fondue, or as dried meat like beef jerky. It may be ground then formed into patties (as hamburgers or croquettes), loaves, or sausages, or used in loose form (as in "sloppy joe" or Bolognese sauce).

Some meat is cured by smoking, which is the process of flavoring, cooking, or preserving food by exposing it to the smoke from burning or smoldering plant materials, most often wood. In Europe, alder is the traditional smoking wood, but oak is more often used now, and beech to a lesser extent. In North America, hickory,mesquite, oak, pecan, alder, maple, and fruit-tree woods are commonly used for smoking. Meat can also be cured by pickling, preserving in salt or brine. Other kinds of meat are marinated and barbecued, or simply boiled, roasted, or fried.

Meat is generally eaten cooked, but there are many traditional recipes that call for raw beef, veal or fish (tartare).Steak tartare is a meat dish made from finely chopped or minced raw beef or horse meat. Meat is often spiced or seasoned, particularly with meat products such as sausages. Meat dishes are usually described by their source (animal and part of body) and method of preparation (e.g., a beef rib). Meat is a typical base for making sandwiches. Popular varieties of sandwich meat include ham, pork, salami and other sausages, and beef, such as steak, roast beef, corned beef, pepperoni, and pastrami. Meat can also be molded or pressed (common for products that include offal, such as haggis and scrapple) and canned.

Issues

Meat is part of the human diet in most cultures. Many people, however, choose not to eat meat (this is referred to as vegetarianism) or any food made from animals (veganism). The reasons for not eating all or some meat may include ethical objections to killing animals for food, health concerns, environmental concerns or religious dietary laws.

Ethics of Eating Meat

Ethical issues regarding the consumption of meat can include objections to the act of killing animals or to the agricultural practices used in meat production. Reasons for objecting to killing animals for consumption may include animal rights, environmental ethics, or an aversion to inflicting pain or harm on other sentient creatures. Some people, while not vegetarians, refuse to eat the flesh of certain animals, such as cats, dogs, horses, or rabbits, due to cultural or religious taboos. In some cases, specific meats (especially from pigs and cows) are forbidden within religious traditions.

Some people eat only the flesh of animals which they believe have not been mistreated, and abstain from the meat of animals reared in factory farms or from particular products such as foie gras and veal. Some people also abstain from milk and its derivatives because the production of veal is a byproduct of the dairy industry. The ethical issues with factory farming relate to the high concentration of animals, animal waste, and the potential for dead animals in a small space. Critics argue that some techniques used in intensive agriculture can be cruel to animals. Foie gras is a food product made of the liver ofducks or geese that has been specially fattened by force feeding them corn. Veal is criticised because the veal calves may be highly restricted in movement; have unsuitable flooring; spend their entire lives indoors; experience prolonged sensory, social, and exploratory deprivation; and are more susceptible to high amounts of stress and disease.

Religious Traditions

The religion of Jainism has always opposed eating meat, and there are also schools of Buddhism, and Hinduism that condemn the eating of meat. Jewishdietary rules (*Kashrut*) allow certain (*kosher*) meat and forbid other (*treif*). Among the numerous laws that form part of *kashrut* are the prohibitions on the consumption of unclean animals (such as pork, shellfish (both Mollusca and Crustacea) and most insects) and mixtures of meat and milk. Similar rules apply in Islamic dietary laws. According to the Quran, the only foods explicitly forbidden are meat from animals that die of themselves, blood, the meat of swine (porcine animals, pigs), and animals dedicated to other than Allah (either undedicated or dedicated to idols).In Sikhism only Kutha meat is forbidden and the proscribed method of killing is Jhatka, however there are sects that oppose eating meat.

Health

A study of 400,000 subjects conducted by the European Prospective Investigation into Cancer and Nutrition and published in 2013 showed "a moderate positive association between processed meat consumption and mortality, in particular due to cardiovascular diseases, but also to cancer."

A 1999 metastudy combined data from five studies from western countries. The metastudy reported mortality ratios, where lower numbers indicated fewer deaths, for fish eaters to be 0.82, vegetarians to be 0.84, occasional meat eaters to be 0.84. Regular meat eaters and vegans shared the highest mortality ratio of 1.00. The study reported the numbers of deaths in each category, and expected error ranges for each ratio, and adjustments made to the data. However, the "lower mortality was due largely to the relatively low prevalence of smoking in these [vegetarian] cohorts".

In response to changing prices as well as health concerns about saturated fat and cholesterol, consumers have altered their consumption of various meats. A USDA report points out that consumption of beef in the United States between 1970–1974 and 1990–1994 dropped by 21%, while consumption ofchicken increased by 90%. During the same period of time, the price of chicken dropped by 14% relative to the price of beef. In 1995 and 1996, beef consumption increased due to higher supplies and lower prices.

Cancer

In recent years, health concerns have been raised about the consumption of meat increasing the risk of cancer. In particular, red meat and processed meat were found to be associated with higher risk of cancers of the lung, esophagus, liver, and colon, among others, although also a reduced risk for some minor type of cancers. Another study found an increase risk of pancreatic cancer for red meat and pork. That study also suggests that fat and saturated fat are not likely contributors to pancreatic cancer. Animal fat, particularly from ruminants, tends to have a higher percentage of saturated fat vs. monounsaturated and polyunsaturated fat when compared to vegetable fats, with the exception of some tropical plant fats; consumption of which has been correlated with various health problems. The saturated fat found in meat has been associated with significantly raised risks of colon cancer,although evidence suggests that risks of prostate cancer are unrelated to animal fat consumption.

However, many research papers do not support significant links between meat consumption and various cancers. Key et al. found that "There were no significant differences between vegetarians and nonvegetarians in mortality from cerebrovascular disease, stomach cancer, colorectal cancer, lung cancer, breast cancer, prostate cancer or all other causes combined." Truswell reviewed numerous studies, concluding that the relationship of colorectal cancer with meat consumption appeared weaker than the "probable" status it had been given by the World Cancer Research Foundation in 1997. A study by Chao et al. (2005) found an apparent association of colorectal cancer with red meat consumption after adjustment for age and energy intake. However, after further adjustment for body mass index, cigarette smoking and other covariates, no association with red meat consumption was found. Alexander conducted a meta-analysis which found no association of colorectal cancer with consumption of animal fat or protein. Based on European data (EPIC-Oxford study), Key et al. found that incidence of colorectal cancer was somewhat lower among meat eaters than among vegetarians. A study within the European Prospective Investigation into Cancer and Nutrition found that association between esophageal cancer risk and total and processed meat intake was not statistically significant. The dissimilar findings indicate that caution is needed in considering claims of dietary links to cancer occurrence.

Heart disease

The correlation of meat consumption to increased risk of heart disease is controversial. Some studies fail to find a link between red meat consumption and heart disease (although the same study found statistically significant correlation between the consumption of processed meat and cancer), while another study, a survey, conducted in 1960, of 25,153 California Seventh-Day Adventists, found that the risk of heart disease is three times greater for 45-64 year old men who eat meat daily, versus those who did not eat meat. A major Harvard University study in 2010 involving over one million people who ate meat found that only processed meat had an adverse risk in relation to coronary heart disease. The study suggests that eating 50g (less than 2oz) of processed meat per day increases risk of coronary heart disease by 42%, and diabetes by 19%. Equivalent levels of fat, including saturated fats, in unprocessed meat (even when eating twice as much per day) did not show any deleterious effects, leading the researchers to suggest that "differences in salt and preservatives, rather than fats, might explain the higher risk of heart disease and diabetes seen with processed meats, but not with unprocessed red meats."

Bacterial Contamination

A 2011 study by the Translational Genomics Research Institute showed that nearly half (47%) of the meat and poultry in U.S. grocery stores were contaminated with *S. aureus*, with more than half (52%) of those bacteria resistant to antibiotics.

Cooking

Meat can transmit certain diseases, but complete cooking and avoiding recontamination reduces this possibility.

Several studies published since 1990 indicate that cooking muscle meat creates heterocyclic amines (HCAs), which are thought to increase cancer risk in humans. Researchers at the National Cancer Institute published results of a study which found that human subjects who ate beef rare or medium-rare had less than one third the risk of stomach cancer than those who ate beef medium-well or well-done. While eating muscle meat raw may be the only way to avoid HCAs fully, the National Cancer Institute states that cooking meat below 212 °F (100 °C) creates "negligible amounts" of HCAs. Also, microwavingmeat before cooking may reduce HCAs by 90%.

Nitrosamines, present in processed and cooked foods, have been noted as being carcinogenic, being linked to colon cancer. Also, toxic compounds called PAHs, or polycyclic aromatic hydrocarbons, present in processed, smoked and cooked foods, are known to be carcinogenic.

Environmental Impact

Various environmental effects are associated with meat production. Among these are greenhouse gas emissions, fossil energy use, water use, water quality changes, and effects on grazed ecosystems. The occurrence, nature and significance of these effects varies among livestock production systems. Grazing of livestock can be beneficial for some wildlife species, but not for others. Targeted grazing of livestock is used as a food-producing alternative to herbicide use in some vegetation management. Meat-producing livestock can provide environmental benefits through waste reduction, e.g. conversion of human-inedible residues of food crops. Manure from meat-producing livestock is used as fertilizer; it may be composted before application to food crops. Substitution of animal manures for synthetic fertilizers in crop production can be environmentally significant, as between 43 and 88 MJ of fossil fuel energy are used per kg of nitrogen in manufacture of synthetic nitrogenous fertilizers.

Imitation Meat

Various forms of imitation meat have been created for people who wish not to eat meat but still want to taste its flavor and texture. Meat imitates are typically some form of processed soybean (tofu, tempeh), but they can also be based on wheat gluten or even fungus (quorn).

Misidentification

With the rise of complex supply chains, including cold chains, in developed economies, the distance between the farmer or fisherman and customer has grown, increasing the possibility for intentional and unintentional misidentification of meat at various points in the supply chain.

In 2013, reports emerged across Europe that products labelled as containing beef actually contained horse meat. In February 2013 a study was published showing that about one-third of raw fish are misidentified across the United States.

ETHICS OF EATING MEAT

In many societies, controversy and debate have arisen over the ethics of eating animals. The fundamental ethical objection to meat eating is that for most people living in the developed world it is not necessary for their survival or health; hence, it is concluded, slaying animals just because people like the taste of meat is wrong and morally unjustifiable. Ethical vegetarians may also object to theagricultural practices underlying the production of meat, or cite concerns about animal welfare, animal rights, environmental ethics, and religious scruples. In response, proponents of meat eating have adduced various scientific, nutritional, cultural and religious arguments in support of the practice. In between, some meat eaters object only to rearing animals in certain ways, such as factory farms; others avoid only certain meats, such as veal or foie gras.

Ethical Views on Eating Meat

Princeton University and University of Melbourne professor and pioneer of the animal liberation movement, Peter Singer, has long argued that if it is possible to survive and be healthy without eating meat, fish, dairy or eggs, one ought to choose that option instead of causing unnecessary harm to animals. In *Animal Liberation*, Singer argued that because nonhuman animals feel, they should be treated according to utilitarian ethics. Singer's work has been widely cited by animal rights campaigners as well as ethical vegetarians and vegans. Ethical vegetarians point out that reasons for not hurting or killing animals are similar to the reasons for not hurting or killing humans. Ethical vegetarians argue that killing an animal, like killing a human, can only be justified in extreme circumstances; consuming a living creature just for its taste, for convenience, or out of habit is not justifiable. Some ethicists have added that humans, unlike other animals, are morally conscious of their behavior and have a choice; this is why there are laws governing human behavior, and why it is subject to moral standards.

When people choose to do things about which they are ambivalent and which they would have difficulty justifying, they experience a state of cognitive dissonance, which can lead to rationalization, denial or even self-deception. For example, a 2011 experiment found that when the harm that their meat-eating causes animals is explicitly brought to people's attention, they tend to rate those animals as possessing fewer mental capacities compared to when the harm is not brought to their attention. This is especially evident when people expect to eat meat in the near future. Such denial makes it less uncomfortable for people to eat animals. The data suggest that people who consume meat go to great lengths to try to resolve these moral inconsistencies between their beliefs and behaviour by adjusting their beliefs about what animals are capable of feeling. Ethical vegetarians argue that it is behaviour rather than supporting beliefs that should be adjusted.

Treatment of Animals

Ethical vegetarian concerns have become more widespread in developed countries particularly because of the spread of (1) factory farming, (2) more open and graphic documentation of what human meat-eating entails for its victims, and (3) environmental consciousness. Some proponents of meat eating argue that the current mass demand for meat has to be satisfied with a mass-production system, regardless of the welfare of animals. Less extreme proponents argue that practices like well-managed free-range rearing and the consumption of hunted animals, particularly from species whose natural predators have been significantly eliminated, could satisfy the demand for mass-produced meat.

Milk and Eggs

One of the main differences between a vegan and a typical vegetarian diet is the avoidance of both eggs and dairy products such as milk, cheese, butter and yogurt. Ethical vegans do not consume dairy or eggs because their production causes animal suffering and/or premature death.

To produce milk from dairy cattle, all calves are separated from their mothers soon after birth and fed milk replacement in order to retain the cows' milk for human consumption.

Animal welfare advocates point out that this breaks the natural mother and calf bond. Unwanted male calves are either slaughtered at birth or sent for veal production. To prolong lactation, dairy cows are almost permanently kept pregnant through artificial insemination.Although cows' natural life expectancy is about twenty years, after about five years the cows' milk production has dropped; they are then considered "spent" and sent to slaughter for meat and leather.

In battery cage and free-range egg production, unwanted male chicks are culled and killed at birth during the process of securing a further generation of egg-laying hens.

Animal Consciousness

Eugene Linden, author of *The Parrot's Lament* suggests that many examples of animal behavior and intelligence seem to indicate emotion, and a level of consciousness that we would normally ascribe only to our own species. Many others have written about animal consciousness

Philosopher Daniel Dennett counters that:

Consciousness requires a certain kind of informational organization that does not seem to be 'hard-wired' in humans, but is instilled by human culture. Moreover, consciousness is not a black-or-white, all-or-nothing type of phenomenon, as is often assumed. The differences between humans and other species are so great that speculations about animal consciousness seem ungrounded. Many authors simply assume that an animal like a bat has a point of view, but there seems to be little interest in exploring the details involved.

Some have argued that sentience (being able to feel sensations, emotions and pain) is not the same as self-awareness (being aware of oneself as an individual). Generally, only the handful of animals that have passed the mirror test are confidently considered to be self-aware.

Peter Singer, in his ethical philosophy of what it is to be a "person" argues that livestock animals feel enough to deserve better treatment than they receive.

Pain

A related argument revolves around non-human organisms' ability to feel pain. If animals can be shown to suffer, as humans do, many of the arguments against human suffering could be extended to animals One such reaction is transmarginal inhibition, a phenomenon observed in humans and some animals akin to mental breakdown.

As noted by John Webster, a professor of animal husbandry at Bristol:

People have assumed that intelligence is linked to the ability to suffer and that because animals have smaller brains they suffer less than humans. That is a pathetic piece of logic, sentient animals have the capacity to experience pleasure and are motivated to seek it, you only have to watch how cows and lambs both seek and enjoy pleasure when they lie with their heads raised to the sun on a perfect English summer's day. Just like humans.

Debate over Animals killed in Crop Harvesting

Steven Davis, a professor of animal science at Oregon State University, argues that the least harm principle does not require giving up all meat. Davis states that a diet containing beef from grass-fed ruminants such as cattle would kill fewer animals than a vegetarian diet, particularly when one takes into account animals killed by agriculture. This conclusion has been criticized by Jason Gaverick Matheny (founder of *in vitro* meat organization New Harvest) because it calculates the number of animals killed per acre (instead of per consumer). He claims that when the numbers are adjusted, Davis' argument shows veganism as perpetrating the least harm. Davis' argument has also been criticized by Andy Lamey for being based on only two studies that may not represent commercial agricultural practices. When differentiating between animals killed by farm machinery and those killed by other animals, he claims the studies again show veganism to do the "least harm".

Jay Bost, agroecologist and winner the New York Times essay contest on the ethics of eating meat, summarized his argument in the following way: "eating meat raised in specific circumstances is ethical; eating meat raised in other circumstances is unethical" in regard to environmental usage. He proposes that if "ethical is defined as living in the most ecologically benign way, then in fairly specific circumstances, of which each eater must educate himself, eating meat is ethical." The specific circumstances he mentions include using animals to cycle nutrients and convert sun to food. Ethicists like Tom Regan and Peter Singer define ethical in terms of suffering rather than ecology.

Environmental Argument

Permaculture is an approach to designing human settlements and agricultural systems that are modeled on the relationships found in natural ecologies. Some people choose to be vegetarian or vegan for environmental reasons.

The use of large industrial monoculture that is common in industrialized agriculture, typically for feed crops such as cornand soy is more damaging to ecosystems than more sustainable farming practices such as organic farming,permaculture, arable, pastoral, and rain-fed agriculture. Other concerns include the wasting of natural resources, such as food, water, etc.

Animals that feed on grain or rely on grazing require more water than grain crops. According to the USDA, growing crops for farm animals requires nearly half of the U.S. water supply and 80% of its agricultural land. Animals raised for food in the U.S. consume 90% of the soy crop, 80% of the corn crop, and 70% of its grain. In tracking food animal production from the feed through to the dinner table, the inefficiencies of meat, milk and eggproduction range from a 4:1 energy input to protein output ratio up to 54:1. The result is that producing animal-based food is typically much less efficient than the harvesting of grains, vegetables, legumes, seeds and fruits, though this might not be true to the same extent for animal husbandry in the developing world where factory farming is almost non existent, making animal-based food much more sustainable.

Some have described unequal treatment of humans and animals as a form of species bias such as anthropocentrism or human-centeredness. Val Plumwood (1993, 1996) has argued that anthropocentrism plays a role in green theory that is analogous to androcentrism in feminist theory andethnocentrism in anti-racist theory. Plumwood calls human-centredness "anthropocentrism" to emphasize this parallel. By analogy with racism and sexism,Melanie Joy has dubbed meat-eating "carnism."

RELIGIOUS TRADITIONS OF EATING MEAT

Hinduism holds vegetarianism as an ideal, for three reasons: the principle of nonviolence (ahimsa) applied to animals; the intention to offer only "pure" (vegetarian) or sattvic food to a deity and then to receive it back as prasad; and the conviction that an insentient diet is beneficial for a healthy body and mind and that non-vegetarian food is detrimental for the mind and for spiritual development. Buddhist vegetarianism has similar strictures against hurting animals. The actual practices of Hindus and Buddhists vary according to their community and according to regional traditions. Jains are especially rigorous about not harming sentient organisms.

Islamic Law and Jewish religious traditions have similar dietary guidelines called Halal and Kashrut. In Judaism, meat that may be consumed according to*halakha* (Jewish law) is termed *kosher*; meat that is not compliant with Jewish law is called *treif*. Causing unnecessary pain to animals is prohibited by the principle of Tza'ar Ba'alei Chayim. While it is neither required nor prohibited for Jews to eat meat, a number of medieval scholars of Judaism, such asJoseph Albo and Isaac Arama, regard vegetarianism as a moral ideal. Rabbi Kook speaks of vegetarianism as an ideal, and points to the fact that Adam and Eve did not partake of the flesh of animals as all humans and animals were originally commanded by God to only eat plants. According to Richard H. Schwartz, president of Jewish Vegetarians of North America and author of the book *Judaism and Vegetarianism*, God's original plan was for mankind to be vegetarian.

In Christianity as practised by members of Eastern Orthodox Church, Roman Catholic Church, Greek Catholic Church and others, it is prohibited to eat meat in times of fasting. Rules of fasting also vary. There are also Christian monastic orders that practice vegetarianism.

2

Principles of Meat Processing Hygiene

Meat processing hygiene is part of Quality Management (QM) of meat plants and refers to the hygienic measures to be taken during the various processing steps in the manufacture of meat products. Regulatory authorities usually provide the compulsory national framework for food/meat hygiene programmes through laws and regulations and monitor the implementation of such laws. At the meat industry level, it is the primary responsibility of individual enterprises to develop and apply efficient meat hygiene programmes specifically adapted to their relevant range of production. Operations in meat processing plants comprise the manufacture of value-added meat products from primary products of meat origin and non-meat origin. There are three principles of meat hygiene, which are crucial for meat processing operations.

- Prevent microbial contamination of *raw materials*, intermediate (*semi-manufactured*) *goods* and *final products* during meat product manufacture through absolute cleanliness of tools, working tables, machines as well as hands and outfits of personnel.
- Minimize microbial growth in *raw materials, semi-manufactured goods* and *final products* by storing them at a low temperature.
- Reduce or eliminate microbial contamination by applying heat treatment at the final processing stage for extension of shelf life of *products* (except dried and fermented final products, which are shelf-stable through low a_w and pH)

The above three principles guide meat hygiene programmes in the further processing of meat. However, meat processing hygiene is more complex. In particular, the hygienic treatment of meat before reaching the processing stage is of utmost importance for the processing quality of the meat. Failures in slaughter hygiene, meat cutting and meat handling/transportation and in the hygiene of by-products and additives will all contribute to quality losses and deterioration of the final processed meat products. Highly contaminated raw meat is unsuitable for further processing. Final products made from hygienically deficient raw meat materials are unattractive in colour, tasteless or untypical in taste with reduced shelf life due to heavy microbial loads. Moreover, there is also the risk of presence of food poisoning microorganisms, which can pose a considerable public health hazard.

In the light of growing consumer consciousness as well as regionalization and globalization in trade, quality conscious meat plants need *internal quality control/quality management schemes* not only for the final products but also for the raw materials and the various processing steps.

Such Quality Management Schemes (QM) have technical and hygienic components. *Technical* aspects encompass *product composition, processing technologies, packaging, storage*and *distribution*. Details on the manufacturing practice for each individual group of meat products are included in the chapters on processing technology. For the*sanitary quality and safety* related to meat processing, two useful schemes can be applied known as

- Good Hygienic Practices (GHP) and
- Hazard Analysis and Critical Control Point (HACCP) Scheme.

Both schemes are not verbally laid down in codes ready to be used for the various purposes in the meat sector although some generic examples can be accessed in handbooks or via internet. Factory and production specific versions need to be established and compiled by taking into account official laws and regulations as well as recommended codes of practice.

PERSONAL HYGIENE

Specific preventive measures to avoid cross contamination (e.g. plant internal transports of raw materials and finished products must not cross each other)

Specific food handling procedures (e.g. meat containers must not directly be placed on the floor, but on stands, pallets etc.)

Suggested control points directly related to meat processing and therefore suited for the establishment of CCPs are:

- unloading bay for raw materials (meat and non-meat ingredients),
- cold storage rooms,
- meat cutting and preparation facilities,
- facility for handling non-meat additives,
- meat comminuting units (grinders, bowl choppers etc.),
- filling equipment and casings,
- heat treatment facilities (smokehouses, cooking vats, autoclaves),

- packaging equipment and materials (including canning),
- cold store for final products,

It is up to the individual meat processing plant to decide, at which points in the processing line CCPs should be established. This will vary from meat plant to meat plant, depending on plant lay-out equipment, type of products and also on previously experienced accidentally occurred shortcomings.

Establishment of Critical Limits for each CCP

Critical limits correspond to the extreme (highest and lowest) values acceptable from the point of view of product safety. This does not always imply that a numerical value has to be fixed. Monitoring may also be based on visual observation, e.g. dirt/faecal contamination of meat, changes to untypical colour, changes in product structure or texture. Besides such sensory parameters, numerical critical limits must be specified for each objective control measure at each CCP. Criteria often used include temperature, time, moisture level, pH, and water activity.

Examples: Visual check of damage to packaged incoming raw materials (rejection in case of severely damaged packages of meat materials or additives)

Visual check of contamination of raw materials (meat, fat). Discolouration (rejection of meat or fat in severe cases), meat potentially contaminated with food poisoning agents (e.g. minimal dirt contamination to be trimmed off, critical dirt or fecal contamination leads to rejection of the meat)

Temperature control of meat derived from slaughterhouses/cutting plants (e.g. d" +4°C)

pH of incoming meat (e.g. < 6.0 for pork, < 5.7 for beef)

Visual check during meat cutting and grading (e.g. to separate and discard unsuitable meat tissues such as those containing parasites, abscesses, etc.) Moisture content expressed as a_w (refers mainly to dry fermented products which should not be packaged or marketed if moisture content keeps above a certain level)

Additives (some products require a certain salt level for better stability in hot environments; nitrite levels should be high enough to inhibit bacterial growth but below toxic levels; the same applies to chemical preservatives)

Control of pasteurization parameters (ensure sufficient cooking, measured as core temperatures in products, e.g. 74°C)

Control of sterilization temperature and time for canned products (e.g. ensure that desired F-values are reached, e.g. F value 4 in fully sterilized canned products)

Visual appearance and texture of final products (greenish discolouration and slimy surfaces as signs of microbial growth, mould growth on surfaces of dried sausages)

Establishment of a Monitoring System for each CCP

Monitoring is the regular/periodic measurement or observation at a CCP to determine whether a critical limit or target level has been met. The monitoring procedure must be able to detect loss of control at the CCP. Monitoring at CCPs should deliver results rapidly in order to enable corrective action during processing. Lengthy analytical testing is not practicable in the context. Hence most of the testing for critical limits listed in (3) is visual, physical and to some extent chemical. The slower microbiological testing does not allow immediate corrective action. Physical and chemical pattern to be instantly measured or monitored in meat processing lines include:

Establishment of Corrective Actions

Corrective actions are those actions to be taken either when monitoring results show that

- a CCP has deviated from its specified critical limit or target level or
- when monitoring results indicate a trend towards loss of control

Action taken must reduce to safe level or eliminate the actual or potential hazard identified.

Corrective actions are for example:

- Reject incoming meat with too high internal temperatures
- Adjust temperature for refrigerated storage and transport of meat
- Remove with clean knives minimal visual contamination of meat surface, reject heavily contaminated meat
- Adjust cooking and sterilization parameters (temperature/time)
- Reject meat with too high pH
- Adjust quantity of curing substances (level of nitrite, nitrite curing salt should contain 99,5% common salt and 0,5% nitrite)
- In case of dry fermented products: If a_w of processed products is too high, stop packaging in water vapour impermeable packages

Products with suspected hygienic deficiencies have to be separated from other products. Additional treatments may have to be applied, e.g. additional heat treatment in case of undercooking. Final judgement (if fit or unfit for consumption) has to be made by responsible, competent persons. Interventions at CCPs are carried out based on instant observation of hygienic failures/ shortcomings. Corrective actions should be documented in the HACCP written records.

Establishment of Verification Procedures

Procedures are needed to ensure that the HACCP system is working correctly. Particular attention must be given to the monitoring frequency, which may be daily or several times a day or more frequently. Checks on the persons doing the monitoring should be done regularly as well as calibration of instruments used.

Established critical limits can be revalidated (changed) in the light of new developments. The system as a whole for individual products has to be reviewed in case of introducingchanges in the processing technology such as changes in raw materials, product composition, processing equipment or packaging systems.

Test results derived from GHP routine quality control, in particular microbiological analysis, are valuable supplementary information within the HACCP system, support the verification process and prove the practicability of HACCP.

Establishment of Documents and Records

Documents and records must be produced commensurate with the nature and size of the food business to demonstrate the application of principles 1-6. These documents serve for the competent authorities to evaluate the efficacy of the HACCP procedure carried out at the plant. Records also help to trace causes of problems that were encountered during past production.

This documentation includes amongst others:

- Certification on receipt of raw meat materials and non-meat ingredients documenting supplier compliance with processor's specifications
- CCP determinations (for each product)
- Critical limits set and results achieved for each CCP (including possible deviations from critical limits and corrective actions)
- Modifications introduced to the system in the light of changes of technology or other developments

HAZARD ANALYSIS CRITICAL CONTROL POINT

Internal sanitary related control and monitoring system in food plants with the aim of preventing/minimizing or eliminating health hazards to consumers. HACCP identifies, evaluates and controls hazards, which are significant for food safety. The characteristics of HACCP are:

- Potential for immediate prevention measures before or during production to counteract suspected or emerging health risks
- Exclusively aimed at health risks to consumers

Food plant internal control procedures based on HACCP principles have become an obligation worldwide in many countries with advanced food industries. HACCP procedures are imposed on relevant food plants by the competent authorities, whose task is to assess and evaluate the correct application and conduct of HACCP. The food plants themselves are responsible for the proper implementation of HACCP, such as monitoring of sensory, physical and chemical parameters during production and immediate intervention in case of emerging health risks and recording of results.

Requirements for introduction of HACCP schemes are yet different from region to region. In a number of countries (e.g. EU, US) meat plants in general have to comply with HACCP, whereby for smaller plants or such specializing in limited activities or products, simplifications or exceptions exist. In some other parts of the world, HACCP schemes are not yet commonly introduced. However, it can be anticipated that such plants involved in regional or global distribution of food will also be obliged to comply with HACCP principles.

Basic elements of HACCP in meat processing plants

- Every single meat product with product specific technology requires a specifically designed individual HACCP scheme.
- As a precondition for implementing HACCP concepts, hazard analysis and risk assessment referring to meat plant specific processing methods or products, have to be carried out.
- Critical control points (CCPs) have to be identified, critical limits be established and monitoring systems properly implemented.

The HACCP scheme is subdivided into seven consecutive steps ("principles"). Through these seven HACCP principles a practical approach is provided to identify potential significant hazards to consumers' health and to take relevant corrective actions:

Hazard Analysis and Risk Assessment

The first principles requires initially the exact description of the products to be fabricated, including product composition, texture/structure, processing details (such as degree of comminuting, additives, filling, heat treatments), packaging and if applicable chemical and microbiological criteria.

Once the characteristics of each product are detailed, potential hazards to consumers' health during processing are identified. Hereunder, a summary listings of hazards are given, from where those hazards likely to be associated with the fabrication of a specific meat product can be identified.

Examples for Hazards in Meat Processing

Biological hazards: Parasites (causing zoonotic diseases), bacteria (causing food poisoning/food borne infections and intoxications), moulds (mycotoxins causing food borne intoxications), viruses (causing food borne infections)

Physical hazards: Rests of unwanted materials (glass, bone fragments, animal teeth/in case of processing head meat, metal fragments such as sausage clips, broken knife blades, needles, plastics, stones)

Chemical hazards: Contaminants (heavy metals, PCB's, chemical solvents, cleaning and disinfection compounds)

Residues (veterinary drugs, feed additives, pesticides)

Food additives with risk of overdoses (nitrate/nitrite, chemical preservatives)

Identification of Critical Control Points (CCP)

A CCP is defined as any point or procedure in a specific food system, where loss of control may result in an unacceptable health risk. CCPs can be located at any point along the production line of a specific meat product, where biological, physical and chemical hazards may occur and where such risks can be controlled and/or eliminated. CCPs should only be established, where firm methods for control and monitoring can be applied.

GOOD HYGIENIC PRACTICES (GHP)

Good Hygienic Practices/GHP follows general hygienic rules and applies recognized hygienic principles as well as laws and regulations issued by the competent authorities, referring to*meat and meat products, equipment, premises* and *personnel*. GHP schemes are *not factory specific*, they apply to all types of meat plants. They are intended to establish and maintain acceptable hygienic standards in relevant meat operations. There is more emphasis on slaughter hygiene in GHP schemes for slaughterhouses and more emphasis on meat processing hygiene in GHP schemes for meat products manufacturing enterprises. However in principle, GHP schemes remain interchangeable for similar types of meat plants.

GHP for meat processing plants refers principally to:

- Appropriate functional plant layout and sanitary design of equipment
- Raw materials that meet hygiene quality standards
- Processing methods that allow safe handling of food
- Appropriate waste and pest control measures
- Appropriate sanitation procedures (cleaning and disinfection)
- Compliance with potable water criteria
- Regular examination of health and personal hygiene of staff
- Regular training of staff on hygiene requirements

Hazard Analysis and Critical Control Point Scheme (HACCP)

HACCP are *factory* and *product specific* strictly sanitary control schemes that shall prevent, detect, control and/or reduce to save levels *accidentally occurring hazards* to consumers' health. Despite GHP in place, accidental hazards cannot be ruled out and may occur at any processing step of the individual meat product. Specifically for *meat processing plants*, such hazards may be provoked by failures such as:

- batches of incoming raw meat materials with abnormal tissues or heavy contamination,
- breakdowns in refrigeration,
- failure in cooking/sterilization operations,
- abnormal pH or a_w in raw or finished products,
- errors in levels of application of curing salts and other additives,
- technical problems in sealing of vacuum packages or cans with the risk of recontamination.

HACCP schemes serve as additional alarm systems in the interest of consumer protection to prevent such problems occuring.

The revolutionary idea of HACCP is to implement control measures that focus on prevention rather than relying on end-product-testing. All relevant possible hazards in the entire production chain, from primary production to consumption of each individual product, must be identified and measures taken for their prevention. In case potential hazards should occur, they can bedetected, contained or eliminated at any stage.

Plant personnel have a key role to play and must be trained in hazard detection and elimination. For practical purposes, those possible hazards may be listed on specific templates for confirmation of presence or absence during routine controls. Specific control mechanisms, in the first place of *physical, chemical and visual nature* (temperature, pH, visual check etc.), are installed at selected control points to detect such potential hazards. These control mechanisms are designed to deliver most results almost instantly and allow immediate intervention during the processing phase of food/meat products.

The need for immediate action within HACCP systems excludes *microbiological control* (of raw materials, semi-fabricated products, tools, equipment, and premises) as a directly applicable control measure. Microbiological control takes hours or days to obtain the results, which does not allow corrective interventions during the usually short manufacturing period. However, this does not mean that microbiological control is worthless for HACCP. Routine microbiological control carried out within the framework of GHP is an extreme helpful tool also for HACCP as its results will demonstrate the efficiency of the HACCP-system. Hygienically acceptable microbiological test results are an indicator of the proper functioning of the meat plant's HACCP scheme.

HACCP is *not* a scheme for the assessment and improvement of the general hygienic status of a meat plant. HACCP is not designed to further raise hygienic standards. Excellent conditions as applicable for GHP-conform plants must already be in place. GHP is a prerequisite requirement for the introduction of HACCP.

The misconception still exists that HACCP is intended to raise levels of general hygiene in meat plants with low hygienic standard. HACCP is not workable where plant layout/structure, equipment and/or processing methods do not comply with good hygienic standards.

One important point to distinguish HACCP from GHP is that GHP describes *process requirements and practices* incl. personal hygiene of staff to ensure safety of food. The individual product is not specifically targeted. Unlike GHP, HACCP always focuses *on the individual product*. As technologies vary from product to product, it is obvious that separate HACCP approaches are required for each category of products.

HYGIENE OF MEAT PROCESSING EQUIPMENT

In production lines in the meat industries equipment and hand-tools should be used, which enable workers to perform all operations according to Good Hygienic Practices. It is the responsibility of the meat plant management to provide adequate equipment for all working places. For equipment manufactures, directives have been issued as to proper design and construction of meat processing equipment. Designs must allow easy and profound cleaning and avoid any accumulation of difficult to remove organic matters.

As a principle in modern meat industries it is commonly accepted that tools and surfaces in contact with meat should be made of food grade stainless steel or synthetic materials.Stainless steel must be used for working tables, meat hooks (at least their parts contact in meat), blades of knives, saws, cleavers and axes.

All parts of machinery in contact with meat, fat, sausage mixes and meat ingredients must be of stainless steel such as frozen meat cutter, grinder, meat mixer and tumbler, meat emulsifier, sausage stuffer, brine injector etc. The bowls of bowl cutters are nowadays also mostly made of stainless steel. All the stainless steel parts must be smooth, easily accessible for cleaning and without hidden spaces, where particles of meat materials may accumulate.

Galvanized steel or food-grade aluminium are useful materials in the meat industries as they are non-corrosive. Those materials should however not be in direct contact with meat, as they are not sufficiently smooth or may release unwanted substances. But they are very suitable materials for overhead rails and supporting structures, working platforms and frames for tables and machinery.

Food grade synthetic materials are used for many types of meat containers and for handles of knives and other hand tools, for cutting boards and some internal parts of meat processing equipment such as washers, parts of valves etc.

HYGIENE OF MEAT PROCESSING PREMISES

Meat processing facilities must meet the following basic hygienic standards in order to ensure and maintain clean and hygienic working conditions:

- Adequate rooms for personnel must be available including sections for changing clothes and for personal hygiene.
- Wall windows must be positioned at a sufficient height from the floors in order to allow profound washing and disinfection of floors and walls. Wall windows for processing plants must be at their lowest part at least 2 m high over floor level. Window frames should be of non-corrosive material e.g. aluminium or similar and must not be painted.
- Walls in all rooms, where meat and by-products are handled, must have smooth and easily washable surfaces up to a minimum height of 2 m in processing plants. Walls should preferably be covered with wall tiles or at least with washable paint.
- Floors in the mentioned sections must be impermeable for water and reasonably smooth for good cleaning, but anti-slip for workers safety. They are usually made of fat-resistant concrete. Additional covering by epoxy substances or floor ceramics are possible.
- In order to facilitate proper cleaning, the junction between floor and walls must be coved, i.e. rounded (not rectangular), which can be achieved by extending the floor concrete up to an height of 10-50 cm alongside the walls. If the concrete layer alongside the wall is sufficiently thick (approx. 10-20 cm), it serves also as shock absorber and protects the walls against damage by transport vehicles, such as trolleys, fork lifts etc. Appropriate coves at wallfloor junctions can also be achieved by using special curved wall tiles.
- All wet rooms must have floor drains, which should be covered by non-corrosive metal plates or grills. The covers should be easily removable for proper cleaning of the drains. Drain sinks must be of the siphon type (anti-smell).
- Provisions must be made to channel waste water from hand-wash facilities, cool room evaporators, tool sterilizers, etc. by means of water pipes or similar directly into effluent drains without contaminating the floor.

- Rooms for meat processing should have sufficient ventilation. Air conditioning is only required in meat cutting/deboning rooms (10 - 12°C).
- Supply systems for electrical wiring and pipes for hot and cold water as well as for compressed air should not hamper cleaning operations and be out of reach of possible dirt contamination. Insulations for hot water pipes must have smooth surfaces and be washable.
- Openings for ventilation must be bird- and insect-proof.

GOOD HYGIENIC PRACTICES IN MEAT PROCESSING

Microbial meat spoilage or food poisoning through meat can be prevented if the microbial load/bacterial contamination, which occurs during slaughtering and meat handling, is kept as low as possible. The key for achieving this is strict meat hygiene including an uninterrupted cold chain throughout the entire meat production and handling chain.

Meat hygiene is a complex field, based on regulations by competent authorities and meat plant internal hygiene programmes, to be supervised by the plant management. Those programmes will only be successful if meat plant staff are familiar with and active in observing basic hygiene requirements. In order to facilitate the application of hygiene requirements, it has proven useful to differentiate between:

a. Personal hygiene
b. Slaughter and meat processing hygiene
c. Hygiene of slaughter and meat processing premises
d. Hygiene of slaughter and meat processing equipment

The topics a-d are of equal significance. Negligence in any of the four areas may give rise to hazards, which can cause economic losses and affect consumers' health.

Some key requirements for meat processing plants are listed below. More detailed hygiene requirements are laid down in national regulations and in international codes, such as FAO/WHO CODEX ALIMENTARIUS Code of Hygienic Practice for Meat (CAC/RCP 58-2005). Guidelines on slaughter hygiene or meat transport and storage hygiene are not included hereunder. However, as meat is the primary material for processed meat products, the application of hygienic practices in slaughterhouses and throughout the cold chain is equally important. Principles of sanitation of premises and equipment are described in a separate chapter.

Principles of personal hygiene

- Wear clean protective clothes
- Washing hands before starting work

- Repeatedly washing hands during work
- No finger rings, watches, bracelets
- Access to production areas with working clothes only
- Cleaning/disinfection of hands/tools/clothes if there was contact with highly contaminated subjects or abnormal animal parts likely to contain pathogens.
- Fresh wounds through knife cuts etc. must be covered by a water tight bandage. Workers with purulent wounds are not allowed to work with meat.
- Strict toilet hygiene must be observed (removal of apron, hand washing and hand disinfection). Toilets must be kept clean and must not have direct access to production areas.
- Periodic medical examination of staff

Basic hygiene of meat processing

- Ideally meat cutting/deboning should be carried out in climatized rooms (approx. + 10°C) with low air humidity. Meat should be brought in progressively and not accumulate on work tables.
- Do not hose down floor and wall areas or equipment next to meat processing operations or final products with a power hose.
- Never take meat pieces, which accidentally had contact with the floor or other contaminated surfaces, back onto working tables or into meat processing machines
- Containers for meat, fat, or semi–or fully processed meat products must not be placed directly on the floor but on hygienic stands, pallets etc.

3

Meat Processing Technologies : Standard Practices

On the other hand, chemical or biochemical processes, which often go together with the technical processes, are also part of meat processing technology such as

- Salting and curing
- Utilization of spices and additives
- Smoking
- Fermentation and drying

These processes are described hereunder and in the following chapters.

Cutting (reducing meat particle size)

There are five methods of mechanical meat cutting for which specialized machinery is used:

Mincing (grinding) of Lean and Fatty Animal Tissues

Larger pieces of soft edible animal tissues can be reduced in size by passing them through meat grinders. Some specially designed grinders can also cut frozen meat, others are equipped with devices to separate "hard" tissues such as tendons and bone particles from the "soft" tissues (minced muscle meat particles).

Chopping Animal Tissues in Emulsifying Machines (continuous process)

The animal tissues to be emulsified must be pre-mixed with all other raw materials, functional ingredients and seasonings and pre-cut using grinders or bowl cutters. Thereafter they are passed through emulsifiers (also called colloid mills) in order to achieve the desired build-up of a very finely chopped or emulsified meat mix.

Chopping Animal Tissues in Bowl Cutter (discontinuous process)

Bowl cutters are used to chop and mix fresh or frozen lean meat, fat (and/ or edible offal, if required) together with water (often used in form of ice), functional ingredients (salt, curing agents, additives) and extenders (fillers and/or binders)

Frozen Meat Cutting

Boneless frozen meat blocks can be cut in slices, cubes or flakes by frozen meat cutters or flakers. The frozen meat particles (2-10 cm) can be directly chopped in bowl cutters without previous thawing thus avoiding drip losses, bacterial growth and discoloration which would happen during thawing. For small operations the manual cutting of frozen meat using cleavers or axes is also possible.

Cutting of Fatty Tissues

Back fat is cut in cubes of 2-4 cm on specialized machines to facilitate the subsequent chopping in cutters/emulsifiers. In small-scale operations this process can be done manually.

Salting / curing

Salting – Salt (sodium chloride NaCl) adds to the taste of the final product. The content of salt in sausages, hams, corned beef and similar products is normally 1.5-3%. Solely common salt is used if the cooked products shall have a greyish or greyish-brown colour as for example steaks, meat balls or "white" sausages.

Besides adding to flavour and taste, salt also is an important functional ingredient in the meat industry, which assists in the extraction of soluble muscle proteins. This property is used forwater binding and texture formation in certain meat products.

The preservation effect, which is microbial inhibition and extension of the shelf-life of meat products by salt in its concentrations used for food (on average 1.5-3% salt), is low. Meat processors should not rely too much on this effect unless it is combined with other preservation methods such as reduction of moisture or heat treatment.

Curing – Consumers associate the majority of processed meat products like hams, bacon, and most sausages with an attractive pink or red colour after heat treatment. However experience shows that meat or meat mixes, after kitchen-style cooking or frying, turn brownish-grey or grey.

In order to achieve the desired red or pink colour, meat or meat mixes are salted with common salt (sodium chloride NaCl), which contains a small quantity of the curing agent sodium nitrite ($NaNO_2$). Sodium nitrite has the ability to react with the red meat pigment to form the heat stable red curing colour. Only very small amounts of the nitrite are needed for this purpose.

Nitrite can be safely used in tiny concentrations for food preservation and colouring purposes. Traces of nitrite are not poisonous. In addition to the reddening effect, they have a number of additional beneficial impacts so that the meat industries widely depend on this substance. Levels of 150 mg/kg in the meat product, which is 0.015%, are normally sufficient.

To reduce the risk of overdosing of nitrite salt, a safe approach is to make nitrite available only in a homogeneous mixture with common salt generally in the proportion 0.5% nitrite and the balance of sodium chloride (99.5%). This mixture is called nitrite curing salt. At a common dosage level of 1.5-3% added to the meat product, the desired salty flavour is achieved and at the same time the small amount of nitrite needed for the curing reaction is also provided. Due to the sensory limits of salt addition (salt contents of 4% are normally not exceeded), the amounts of nitrite are kept low accordingly.

A great deal of research has been done with regard to the utilization of nitrite and it can be said that nitrite in meat products is safe if basic rules are adhered to. Nitrite is now recognized a substance with multifunctional beneficial properties in meat processing:

- The primary purpose of nitrite is to create a heat resistant red colour in a chemical reaction with the muscle pigment, which makes cured meat products attractive for consumers.
- Nitrite has a certain inhibitory effect on the growth of bacteria. This effect is particularly pronounced in canned meat products which are usually stored without refrigeration, where small numbers of heat resistant bacteria may have survived but their growth is inhibited by the presence of nitrite.
- Nitrite has the potential of attributing a specific desirable curing flavour to cured products.
- In the presence of nitrite fats are stabilized and rancidity in meat products retarded i.e., an antioxidant effect.

Many attempts have been made to replace nitrite by other substances, which would bring about the same beneficial effects as listed above. Up to now no alternative substance has been found. As the above desirable effects are achieved with extremely low levels of nitrite, the substance can be considered safe from the health point of view. Currently the known advantages of nitrite outweigh the known risks.

Curing of Chopped/comminuted Meat Mixtures

Curing is applied for most chopped meat mixtures or sausage mixes for which a reddish colour is desired. The curing agent nitrite is added in dry form as nitrite curing salt. The reaction of nitrite with the red meat pigment starts immediately. Due to homogenous blending the meat pigments have instant contact with the nitrite. Higher temperatures during processing, e.g. "reddening" of raw-cooked type sausages at 50°C or scalding/cooking of other products at 70-80°C, accelerate the process.

Another accelerating or "catalytic" effect is the addition of ascorbic acid, which slightly lowers the pH of the meat mixture. However, the dosage of ascorbic acid must be low (0.05%), just to provide the slightly acid conditions for the reduction of $NaNO_2$ to NO. A pronounced reduction of the pH would negatively affect the water binding capacity of the product which is not desirable.

PROCESSED MEAT PRODUCTS

Common Cured Meat Products

The commonest cured products are sausages, bacon, pork shoulder, ham, luncheon meat; any type of meat can be cured either as whole cuts or after comminution. Bacon is cured pork, in various countries traditionally made from specified parts of the pig but it can be made from any part. There are modifications of the process including so called sweet cure with added sugar (0.25 per cent) and mild cure with less salt.

Ham is the cured product of the upper leg and buttock of the pig and differs from gammon only in that the latter is cut from the side of bacon after it has been cured. It is stable when raw after a certain period of maturation but is often cooked to pasteurisation temperature, 70°C, or it may be canned at pasteurising temperature. It may be smoked as an additional means of preservation and flavouring. Typical analysis of canned ham per 100 g: 65-72 g water, 18 g protein, 5-12 g fat, 0.5-0.8 MJ, 1100-1250 mg sodium, 1.2-2.7 mg iron, 0.2 mg copper, 2 mg zinc, 0.5 mg thiamin, 0.2-0.25 mg riboflavin, 4 mg niacin, 0.2 mg vitamin B6, and may have residual ascorbic acid 10-60 ma.

Sausages

There are some 800 types of sausage made of comminuted or chopped meat of various kinds, seasoned with salt and spices, often mixed with cereal and packed into natural casings (consisting of the connective and muscle tissue of animal intestines) or made of cellulose, collagen or synthetic materials. There are six main types of sausage - fresh, smoked, cooked, smoked and cooked, semi-dry and dry.

Frankfurters, Bologna, Polish and Berliner sausages are generally made from beef, pork and pork fat comminuted with the addition of curing salts and are smoked and cooked. Thuringer, soft salami, mortadella, and soft cervelat are cooked and semi-dry; pepperoni, chorizos, dry salami and dry cervelat are slowly dried to a hard texture without cooking. The nutrient content of a number of products. There are several variants of each type of sausage; for example, a US table of food composition (Watt and Merrill 1975) includes six types of Frankfurters, namely all meat, with non-fat dried milk with cereal and with both these additions, also raw and canned.

These differences can be attributed to variations in composition and method of analysis, and serve to illustrate the approximate nature of any tables of nutrient contents of processed meat products.

A major reason for the difference in thiamin content in some comminuted meat products between US and UK tables, is the use of a preservative in UK, namely sulphur dioxide, which destroys most of the thiamin.

Liver sausage contains 10-20 per cent liver and in many cases other edible offals. Blood sausage contains 10-20 per cent whole blood with nitrite salt (not precooked). Other components are precooked meat, edible offals, fatty tissue (cooked sufficiently to separate fat with a low melting point) and pigskin. This type of sausage has a firm consistency due to swollen connective tissue and gelatinised collagen.

Fermented sausages are dry sausages including salami, dry pork and beef sausages and summer sausages, that have been subjected to bacterial fermentation. Meat from a variety of animals may be used, including camel, donkey and horse but rarely mutton, goat or venison. Only well-chilled or frozen meat is used and a temperature of -2 to +5°C maintained during chopping to facilitate comminuting of lean and fatty tissues to the particle size desired and to avoid deposition of fat drops in the batter. Added salt prevents the growth of unwanted micro-organisms and extracts salt-soluble proteins to form a protein gel which binds the pieces of meat together. The bacteria originate from the natural flora of the meat and the environment although starter cultures of Micrococcus, Pediococcus cerevisiae, etc., are sometimes used.

During the slow, prolonged fermentation the pH falls to between 4.8 and 5.4 then the product is dried and may be smoked. Fermented sausage is not cooked and preservation depends on the high acidity and high salt content toegther with the low water content.

Additives

Comminuted products such as sausages and luncheon meats are based on lean meat, which, technologically, provides water-holding and meat-binding capacity, with the addition of fatty meats and, sometimes, organ meats. The amount of these is limited otherwise the products have an unattractive soft texture and high shrinkage on cooking.

The ingredients include cereals and potato starch, termed fillers, which also serve to bulk out the supply of meat products ("meat extenders"). Other ingredients include a number of substances which have considerable water-holding and binding capacity ("binders"). These include egg or egg yolk blood plasma, skim milk powder, caseinates, soya isolates, wheat gluten, whey protein and dehydrated products derived from various vegetable proteins (soybean, safflower, corn, peanut and pea protein) and their binding properties depend on their ability to form irreversible gels on mild heating which serve to hold together the small pieces of meat. It is not posssible to generalise about the nutritional value of products of such variable composition.

Other Comminuted Products

These are made from chopped (minced, ground) meat and fat and usually include meat from various parts of the animal including trimmings. In specific products organ meats are also used. Some tissues can be included only in restricted amounts because their texture can adversely affect the product.

Other ingredients are included for the purposes of ekeing out meat supplies or for their capacity to bind the minced pieces and seasoning.

Seasoning is a comprehensive term for ingredients intended to improve flavour such as salt, pepper, spices, herbs and vegetables. Spices commonly used include cinnamon, cassia, clove, ginger, mace, nutmeg, paprika, cardamom, coriander and mustard; herbs include sage, savory, bay leaves, thyme and rosemary; onions and garlic are also used.

Liver and blood have a pronounced colour and flavour and can be included in comminuted meat products only in limited amounts (15-50 per cent of the total) depending on the local acceptability or may be used in specific liver or blood sausages.

Pates can be included with comminuted products since they are made from coarsely or finely chopped meats, precooked and seasoned, and some are cured with salt, nitrite and phosphate.

Luncheon Meat

This is the name given to several products made from finely chopped meat and fat with the addition of cereal and water, preserved by the addition of salt and nitrite and by heating.

Although the nutrient content is very variable a typical analysis of a canned product of this type, per 100 g is:- 5 g water, 13 g protein, 27 g fat (10 g saturated, 12g monounsaturated, 3 g polyunsaturated, 70 g cholesterol),

1.3 MJ, 1000 mg sodium, 1 mg iron, 0.3 mg copper, 2 mg zinc, 0.07 mg B1, 1.8 mg nicotinic acid, 1 mg B12.

Corned Beef

This was originally a by-product from the manufacture of meat extract when it was the only meat product that could be shipped from South America before the introduction of refrigerated transport. The latter is made by hot water extraction of the low quality meat from animals that have taken several years on relatively poor pasture to reach suitable size so is relatively rich in connective tissue (as well as in muscle extractives). The meat is coarsely chopped and immersed in hot water to extract the solubles; the exhausted meat is cured by the addition of coarse grains of salt ("corns") and nitrite and canned, often with the addition of fat. It has an extremely long shelf- life and used to play a major role in military rations and in expeditions.

Other meats such as mutton can be treated similarly and some modern processes use unextracted beef. In the United States the term "corned beef" is applied to what is elsewhere termed "salt beef" *i.e.,* cured whole beef. Since corned beef is made from extracted meat it is low in water-soluble vitamins, containing per 100 g only a trace of thiamin, 2.5 mg nicotinic acid, 0.2 mg riboflavin, 2 mg B12, 3 mg iron, 0.25 mg copper, 6 mg zinc, together with 950 mg sodium and 60 g water, 26 g protein, 12 g fat and supplying 0.9 Ml (220 kcal).

Burgers

Patties made from minced meat have become popular world-wide through the agency of international fast-food outlets. These were originally hamburgers made from beef with the name derived from the Hamburger sausage but may be made from any meat and "muttonburgers" are marketed in regions where beef is not eaten.

The meat content varies from 100 per cent including about 20 per cent fat, to 80 per cent or less with various additions of cereals, onion and water. Since the pattie is raw it is stored frozen.

Typical composition per 100 g raw hamburgers made from 90 per cent meat is;- 56 g water, 15 g protein, 20 g fat (10 g saturated, 9 g monounsaturated, 1 g polyunsaturated, 100 mg cholesterol), 1.1 Ml, 600 mg sodium, 2.5 mg iron, 0.25 mg copper, 3 mg zinc, 0.04 mg B1, 0.2 mg B2, 4 mg nicotinic acid, 0.2 mg B6, 1 ug B12, 1 mg folate, 0.4 mg pantothenate.

Frying causes a small loss of water, about 5 per cent, and a greater loss of fat depending on the method used - whether under a grill, on a heated surface or over a direct flame. Other nutrients, including protein, are proportionately increased.

RECENT DEVELOPMENTS IN MEAT PROCESSING

Mechanically Recovered Meat (MRM)

Also termed mechanically deboned and mechanically separated meat. When bones are trimmed of adhering meat by hand some of the meat is left attached to the bones. It is said that some 2 million tonnes of meat world -wide are discarded with the bone. This can be recovered by a mechanical scraping process in which bones with adhering meat are generally broken into small pieces and the soft components, meat, fat, etc., forced through perforated screens and recovered as a finely comminuted product consisting of a gel of meat, fat and sometimes marrow from the inside of the bones together with fine particles of bone. The bone content should be reduced to a minimum; it has been recommended that the calcium content of MRM expressed on a dry matter basis should not exceed 1.5 per cent.

MRM can increase the overall profitability of meat production not only by increasing meat yield but also by providing a product with emulsifying properties useful in preparing comminuted meat products. Generally MRM is less acceptable than ordinary muscle meat because it is darker in colour due to its high content of haeme protein, and more readily oxidised because of the high fat content. Because of its physical state and the longer process MRM is subject to heavy microbial contamination; if not used immediately it needs freezing to prevent decomposition.

Reformed Meat Products

Apart from the high regard for certain cuts of meat in some areas there is a growing demand for lean rather than fatty cuts. The high cost of such products has led to the use of other, less expensive cuts of meat which are higher in content of connective tissue but can have some of this tissue, together with skin and gristle, removed in a process of comminution and reformation. The pieces of meat are treated in a tumbling device with brine which penetrates the tissues and extracts some of the myofibrillar protein to form a gel. At a low temperature the product can be compressed into thin flakes which bind together when the meat is heated, sometimes with the addition of binding agents such as wheat gluten, egg albumin and soya; the final product can be shaped or sliced. The nutritive value of reformed meat products depends on the tissue used and the added ingredients, and approximates to that of comminuted products.

Protein Extraction

A recently discovered but highly technical process for recovering edible foodstuffs from by-products of low acceptability (such as lungs and stomachs) is extraction of the proteins. The procedure is similar to that of preparing textured vegetable proteins usually made to simulate meat. The proteins are extracted in alkali and reprecipitated from solution by acidifying. If the alkaline solution is forced through fine tubes (spinnerets) into the acid solution it is precipitated in the form of long fibres which can be pressed and texturised to a solid form. High temperature extrusion is also used. Other materials such as soya can be incorporated. There is an added advantage in that the number of micro-organisms is greatly reduced during the process.

A recent method of using low quality meats is the application of the Japanese fish surimi process to meat and poultry by-products. This consists of mechanically deboning and washing to remove water-soluble proteins, enzymes, minerals and fat to leave a product that is stable and with a high concentration of salt-soluble myofibrillar proteins that have the capacity to form strong elastic gels when gently heated. Fish surimi is used to make a range of products by extrusion and shaping and it seems possible to apply the procedure to meat. Low-quality meat is rich in fat and the process is intended to remove the greater part of the fat and to make novel meat products or to add to traditional meat products such as sausages, sliced cooked meat and burgers or other foods.

Curing of Meat

Salt is the only essential ingredient for curing. It retards spoilage by drawing water out of the meat while also killing decay-causing microorganisms. Meat cured with salt alone will store well (but will be tough and dry). An early salt cure was as simple as storing slabs of meat in a barrel of salt. 'Kosher salt' can be used for curing meat and is a type of coarse salt which is usually made without additives. 'Curing salt' is most commonly used today and is a special blend of salt and other ingredients including sodium nitrite (which has become somewhat controversial regarding potential health issues with large doses).

Smoking of Meat

Curing is the first step in the smoking process, essential for good flavour. 'Cold smoking' is best for preservation (and for adding flavour) with temperatures between 70 and 90 degrees F, and is more easily accomplished during cooler months. The cold smoking process may range from 1 to 14 days. The preservative benefit of smoking is that the smoke contains tar-like substances that are deposited on the food. To a greater or lesser extent, they seal the surface, keeping air from coming in contact with the food. Fats in the food will not turn rancid from exposure to air, so smoking is particularly useful for preserving fatty foods. The smoke also kills bacteria. Smoked meats may still spoil quickly though, depending, so refer to research of your particular meat, recipes, and shelf-life storage.

Types of Montreal Smoked Meat

Along with bagels, smoked meat has been popular in Montreal since the nineteenth century, and has taken such strong root in that city that many Montrealers, and even many non-Montrealers, identify it as emblematic of the city's cuisine. Current and former residents and tourists make a point of visiting Montreal's best-known smoked meat establishments such as Schwartz's, Lester's Deli, Roi du Smoked Meat (Smoked Meat King), Abie's, Snowdon Deli, Smoked Meat Pete's, Main's Deli, even taking whole briskets away as take-out. Warm Montreal smoked meat is always sliced by hand in order to maintain its form since doing so with a meat slicer would cause the tender meat to disintegrate. Whole briskets are kept steaming and sliced up on demand when ordered in the restaurant.

Pastrami

Famous among early smokers of meat are the Ashkenazi Jewish communities in Europe, and is often associated with other foods popularised by Jewish communities, such as bagels. In The United States it is referred to as pastrami which is derived from theYiddish pronounced pastrómeh. However, lovers of pastrami vociferously argue that the consistency, flavouring, seasoning and colour of pastrami differs significantly from that of smoked meat. Pastrami comes in two flavours-"old fashioned" which is a process where the meat is naturally aged or cured and "regular" a process whereby additives are used to age the meat. Generally, those who have tested traditional "New York Deli" pastrami agree that it is similar, but not identical, to pastrami Smoked Meat. Both the dish and the word were brought to North America with the wave of Jewish immigration from Bessarabia and Romaniain the second half of the 19th century; it is similar to roast brisket, a signature dish of the local Jewish cuisine of these regions. Smoked meat, also known as salt beef in London, is cured, spiced, and flavoured in ways similar to corned beef. Difference in meat cut and spicing mean that smoked meat's taste is different from either of these, and even varies among recipes.

Smoked Meat in Sikkim, India

Smoke meats from the southern United States, also known as Barbecue or BBQ were developed to use the "cheaper" cuts of meats, as they were typically tough and deemed "undesirable". The "low and slow" cooking methods tenderised the connective tissues in the meat made these cuts a sought after delicacy. Usually done in a "low temp" environment (200 °F to

300 °F) (93 °C to 149 °C), they take a significant amount of time to prepare, which breaks down the connective tissues and collagens within the meat, and renders out of the fat. A sub-category is the plethora of sauces and "mops" that are used to flavour and further tenderise the meats. BBQ has become a culinary art form, and hundreds of competitions are held yearly.

Smoking Reasons

Smoking offers many improvements for meat. Besides enhancing the taste and look, it also increases its longevity, and helps preserve the meat by slowing down the spoilage of fat and growth of bacteria. Smoking meat longer leads to more water loss, and results in a saltier and drier product, which naturally increases its shelf life.Man discovered that in addition to salting and curing meat with nitrates, smoking was a very effective tool in preserving meats.

The advantages of smoking meat are numerous. Smoking:

- Kills certain bacteria and slows down the growth of others.
- Prevents fats from developing a rancid taste.
- Prevents mold from forming on fermented sausages.
- Extends shelf life of the product.
- Improves the taste and flavour.
- Changes the colour; smoked meats shine and simply look better.

Smoked fish develops a beautiful golden colour. The meat on the outside becomes a light brown, red, or almost black depending on the type of wood used, heating temperatures, and total time smoking.

The smell in an ethnic meat store specialising in smoked products can be overwhelming. This experience is not shared with our supermarkets since their products are rarely properly smoked and they are vacuum-sealed to prolong shelf life. Certain classical sausages are smoked for up to 3 days and in today's era it is hard to imagine a manufacturer that will do that. To survive the frantic pace of today's market, water is pumped into the meat, chemicals are added for aesthetic and preservation reasons, and smoking is virtually eliminated by adding liquid smoke. As long as the ingredients are not on the list of chemicals that present danger to us, the Food and Drug Administration does not care what goes into the meat. Taste plays a secondary role, as long as the price is good people will buy the product and supermarkets will keep renewing orders. Smoking to preserve meat's keeping qualities is of less importance today because we can keep the product in a refrigerator or almost indefinitely in a freezer. Originally, curing and smoking was used solely for preservation purposes; today it's done for the love of its flavour.

Smoking may or may not be followed by cooking. Generally we may say that smoking consists of two steps:

1. *Smoking:* Meats are usually cured before they are submitted to smoking.
2. *Cooking:* This step determines the design and quality of your smokehouse as it needs temperature controls, a reliable heat supply and good insulation to hold the temperature when the weather gets cold. If cooking is performed outside the smokehouse, the unit can be incredibly simple, for example an empty cardboard box.

After smoking is done we increase the temperature to about 170° F (76° C) to start cooking. The smoked meats must be cooked to 154° F (68° C) internal temperature and here the quality and insulation of the smoker plays an important role. Nevertheless, the main smoking process is performed below 160° F (71° C).

We know now that the smoked meat must be cooked, but does that mean that it must be cooked inside of the smokehouse? Don't we have wonderfully designed and factory built electrical or gas stoves inside every kitchen? They are insulated, have built-in temperature controls and are almost begging for these smoked sausages to be baked inside. How about putting your smoked meats into a pot full of hot water and cooking these products on top of the stove?

Traditionally smoked meats come almost always from cured parts of pork. The most popular large cuts used for smoking are ham, bacon, butt, loin, back fat and smaller parts such as hocks and jowls. Ribs are normally barbecued. Due to their large size those popular cuts require longer curing times although those times can be somewhat shortened when needle pumping precedes the common wet curing method. Hams can be dry or wet cured, butts and loins are normally wet cured and bacon and back fat are commonly dry cured. Trimmings end up for making sausages.

Smoking Without Nitrates

For those who smoke meats without cures, it will be advisable to smoke them at temperatures well above the danger zone (>160° F, 72° C). Such a product will not be pink but will exhibit a typical grayish colour of cooked meat. Adding cure to meats that will be smoked brings many benefits, one of them is preventing the danger of contracting food poisoning, known as botulism. Barbecued meats are smoked at much higher temperatures which eliminates the danger of Clostridium botulinum producing toxins.

Those who insist on smoking meats without nitrates, should be aware that the internal meat temperature trails the temperature of the smokehouse by about 25° F and to be on the outside of the danger zone, the smoking must be performed at temperatures higher than 170° F (77° C) which in our opinion becomes cooking with smoke. Clostridium botulinum bacteria need moisture, warm temperatures and the absence of oxygen. These are prevalent conditions in a small self contained smoker, where incoming air is kept at minimum in order for the sawdust to smolder and not to burst into the flames. A large outside smokehouse with a separate fire pit is at a smaller risk as there is an ample flow of fresh air that enters smoking chamber together with the smoke. Using dry wood increases safety as less moisture will be created.

Smoking Temperatures

Smoking temperature is one of the most important factors in deciding quality. There is no steadfast rule that dictates exact temperature ranges for different types of smoking. A few degrees one way or the other should not create any problem as long as the hot smoking upper temperature limit is not crossed. Crossing this limit will significantly affect the look and the taste of the product. When smoking, the inside temperature of the smoker cannot exceed 170° F (78° C) for any extended time. At this temperature, fat starts to melt quickly. Once it melts, the sausage inside will be a mass of bread crumbs, have a greasy outside, will lose its shine, and will have an inferior taste. If your sausage:

- Is greasy on the outside.
- Contains spots of grease under the sausage.
- Is too shriveled and wrinkled.
- Has lost its shine and looks opaque.
- Is crumbly inside with little empty pockets.

It means that the internal temperature of the sausage was too high during smoking or cooking. The fats start to melt at very low temperatures and we don't want them to boil and leak through the casings. When faced with excessive temperatures, they begin to melt, and there is no way to undo the damage.

EQUIPMENT USED IN MEAT PROCESSING

In modern meat processing, most of the processing steps can be mechanized. In fact, modern meat processing would not be possible without the utilization of specialized equipment. Such equipment is available for small-scale, medium-sized or large-scale operations. The major items of meat processing equipment needed to fabricate the most commonly known meat products are listed and briefly described hereunder.

MEAT GRINDER (MINCER)

A meat grinder is a machine used to force meat or meat trimmings by means of a feeding worm (auger) under pressure through a horizontally mounted cylinder (barrel). At the end of the barrel there is a cutting system consisting of star-shaped knives rotating with the feeding worm and stationary perforated discs (grinding plates). The perforations of the grinding plates normally range from 1 to 13mm. The meat is compressed by the rotating feeding auger, pushed through the cutting system and extrudes through the holes in the grinding plates after being cut by the revolving star knives. Simple equipment has only one star knife and grinder plate, but normally a series of plates and rotary knives is used. The degree of mincing is determined by the size of the holes in the last grinding plate. If frozen meat and meat rich in connective tissue is to be minced to small particles, it should be minced first through a coarse disc followed by a second operation to the desired size. Two different types of cutting systems are available, the "Enterprise System" and the "Unger System":

- The "Enterprise System" is mainly used in smaller meat grinders with orifice diameters up to 98 mm and consists of one star knife, sharpened only on the side facing the disc, and one grinder plate. Hole diameters can vary from 13 to 5 mm.
- The "Unger System" is used in meat grinders with orifice diameters up to 440 mm and consists of the kidney plate, one or two star knives sharpened on both edges and one or two grinder plates. For a final particle size above 8 mm the recommended setting is kidney plate – star knife – grinder plate. For a final particle size <8 mm the recommended setting is kidney plate – star knife – grinder plate (13 mm) – star knife – grinder plate (6 to 1 mm).

The smallest type of meat grinder is the manual grinder designed as a simple stuffing grinder, i.e. meat material is manually stuffed into the feeder. For all these small machines the Enterprise cutting system is used with one star knife and one grinder plate. These machines are very common everywhere in food processing but their throughput and production capacity is limited due to the small size and manual operation.

The intermediate size meat grinder, also designed as a stuffing grinder, has orifice diameters up to 98 mm. It is driven by a built-in single-phase electrical motor (250 V) and available as both a table and floor model. The meat is put onto the tray and continuously fed by hand into a vertical cylindrical hole leading to the feed auger. The meat or fat is forced by its own weight into the barrel with the rotating feed auger. This type of meat grinder is the most suitable for commercial small-scale operations. Some brands use the Enterprise cutting system, others the Unger system.

Large industrial meat grinders are driven by a three-phase electrical motor (400 V) and equipped with the Unger cutting system. The orifice cylinder diameter of this type of grinder ranges from 114 - 400 mm. Industrial grinders are either designed as stuffing grinders with either tray or hopper or as an automatic mixing grinder. The automatic mixing grinder has a big hopper and the meat falls automatically onto the mixing blades and the feeding worm (auger). The mixing blades and feeding worm can be operated independently with mixing blades rotating in both directions but the feeding worm only towards the cutting set. Most of the industrial meat grinders are also equipped with a device for separating tendons, bone particles and cartilage.

BOWL CUTTER (BOWL CHOPPER)

The bowl cutter is the commonly used meat chopping equipment designed to produce small or very small ("finely comminuted") lean meat and fat particles. Bowl cutters consist of a horizontally revolving bowl and a set of curved knives rotating vertically on a horizontal axle at high speeds of up to 5,000 rpm. Many types and sizes exist with bowl volumes ranging from 10 to 2000 litres. The most useful size for small- to medium-size processing is 20 to 60 litres. In bigger models bowl and knife speed can be regulated by changing gears. Bowl cutters are equipped with a strong cover. This lid protects against accidents and its design plays a crucial role in the efficiency of the chopping process by routing the mixture flow. Number, shape, arrangement, and speed of knives are the main factors determining the performance of the cutter. Bowl cutters should be equipped with a thermometer displaying the temperature of the meat mixture in the bowl during chopping.

Modern large scale bowl cutters may have devices to operate under a vacuum, which helps to improve colour and texture of the meat products by keeping oxygen out of the meat mixes and avoid air pockets. Cutter knives should be adjusted to a distance of 1-2 mm from the bowl for optimal cutting (check the manufacturers recommendations for each model). Most of the large and high-speed bowl cutters are equipped with mechanical discharger devices for emptying the cutter. The process of chopping in a bowl cutter is used for producing fine comminuted products such as frankfurters, bologna, liver sausage etc., and enables processors to offer a much wider range of products.

Filling Machine ("sausage stuffer")

These machines are used for filling all types of meat batter in containers such as casings, glass jars, cans etc. The most common type of filling machine in small and medium size operations is the piston type. A piston is moved inside a cylinder forcing the meat material through the filling nozzle (funnel, stuffing horn) into the containers. Piston stuffers are either attached to the filling table or designed as floor models. In small-scale operations manual stuffers are usually sufficient, sometimes even simple hand-held funnels are used to push meat mixes into casings.

Modern filling machines for larger operations are designed as continuous vacuum stuffers. During the filling process a substantial part of the enclosed air is removed from the product, which helps to improve colour and texture of the finished products. These models are usually equipped with a portioning and twisting devise and have a casing grip devise attached for filling of "shirred" (folded) uncut collagen and plastic casings. This type of continuous filling equipment is relatively expensive and are thus not used in small- to medium-size operations.

Clipping Machine

Clipping machines place small aluminium sealing clips on the sausage ends and replace the manual tying of sausages. They can be used for artificial or natural casings. Clipping machines can also be connected to filling machines. Such machines work with so called casing brakes, which are devices for slow release of the shirred casings from the filling horns ensuring tight filling. Then the filled casing segments are clipped in portions. So called double clipping machines place two clips next to each other, which ensures that the individual sausage portions remain clipped on both ends and easy separation of the sausage portions is possible. When using shirred casings, the time consuming loading of pre-cut casings is no longer necessary. Wastage of casings can be reduced to a minimum by tight filling and leaving only as much casing for the sausage end as needed for the placing of the clips.

Clipping machines are mainly used in larger operations and in most cases operated by compressed air. For medium-scale operations manually operated hand clippers are available.

Smokehouses

Simple smokehouses are used for smoking only. In traditional and small-scale operations the most common methods of smoke generation include burning damp hardwood sawdust, heating dry sawdust or heating pieces of log. But technological progress has changed the smoke generation and application techniques. Methods used in modern meat processing include the following:

Burning/smouldering of Saw Dust

In modern smokehouses (1), smoke generation takes place outside the smoking chamber in special smoke generators with electrical or gas ignition (4). Separate smoke generators allow better control of the quantity and temperature of the smoke produced. The sawdust or chip material (3) is moved from the receptacle to the burning zone (4) by a stirrer or shaker (3). It is ignited by means of an electrically heated plate or by gas flame. A smoke stripper, which is basically a cold water spray, can be placed in the initial part of the smoke pipe and serves to increase the purity of the smoke as undesirable substances are washed out. Smoke with a high degree of desirable smoke components can be obtained in the low temperature range of thermal destruction of saw dust beginning at around 230°C and not exceeding 400°C. The smoke is conveyed directly from the generator to the smoking chamber via a smoke pipe (2). The burned sawdust is collected at the bottom (5).

SMOKE GENERATION THROUGH FRICTION

Timber (3), which is pressed (1) against a fast-rotating steel drum (4) results in pyrolysis of the wood in the favourable temperature range of 300°C to 400°C. The flameless, light, dense and aromatic smoke contains a large proportion of desirable smoking substances and a low proportion of tars. The smoke is conveyed (2) into the smoking chamber. The creation of smoke can be commenced and completed in a matter of seconds. The operation of this type of smoke generators is usually carried out in a discontinuous manner. The smoke quantity and quality can be regulated by changing the speed and time of rotation. As this type of smoke can be produced at relatively low temperatures, it does not carry high amounts of hazardous substances such as benzopyrene.

Smoke Generation through Steam

Overheated steam (3) at approximately 300°C is injected into a compact layer of sawdust (4), which causes thermal destruction of the wood and smoke is generated. This method allows the control of smoke generation temperature by choosing the adequate steam temperature. Impurities in the smoke caused by particles of tar or ash are minimal. The steam- smoke mixture condensates extremely quickly and intensively on the surface and inside the sausage products and produces the desired smoking colour and flavour. No connection to the chimney is required as smoke particles not entering the products settle down in the condensing steam. The condensed water is conducted to the effluent system. Other details of the system are: Hopper and conveyer for sawdust (1,2), smoke duct to smoking chamber (5), ashes (6).

Combined Equipment

Modern facilities can combine smoking, cooking and cooling operations for meat products in one continuous process. By means of automatic stirring systems processing parameters such as smoke generation, temperature (up to 100°C) and relative humidity (up to 100%) required to dry, smoke, or steam-cook any type of product, can be pre-set. With additional refrigerated units installed in the smokehouse, it is also possible to use it as a fermenting/

ripening room for the first crucial steps in production of fermented sausages or raw ham products, where air temperature and air humidity have to be accurately controlled.

Brine Injector

This equipment serves for the injection of brine into meat. Brine is water containing dissolved salt and curing substances (nitrite) as well as additives such as phosphates, spices, sugar, carrageenan and/or soy proteins. The injection is done by introducing pointed needles into the muscle tissue. Brine injection is mainly used for the various types of ham, bacon and other whole muscle products.

Brine injectors are available in different sizes from manually operated single-needle devices for small-scale operations to semi-automated brine injectors with up to 32 needles and more. In large machines the quantity of brine injected into the fresh meat can be determined by pre-setting of pressure and speed. It is very important that all parts of the brine injectors are thoroughly cleaned after every working session and disinfected regularly. Before the injector is used again all hoses and needles should be rinsed with warm water as particles left in the system can block the needles. Absolute cleanliness is necessary as microorganisms remaining in the system would be injected deep into the meat pieces during the operation.

Tumbler or Massager

Tumblers are used for the processing of meat products such as whole-muscle or reconstituted hams. Such machines resemble in principle a drum concrete mixer. A rotating drum with steel paddles inside slowly moves the meat pieces thus causing a mechanical massaging effect. This mechanical process is assisted by the addition of salt and phosphates to achieve equal brine distribution and liberates muscular protein from the meat tissue (protein extraction). The semi-liquid protein substances join the meat pieces firmly together during later heat treatment. For hygienic reasons it is important to place the tumbler below 10°C to avoid excessive microbial growth during lengthy tumbling times (more then 4 hours or even over night). In specific cases it is recommended that the tumbler should be operated refrigerated or inside a cold room below -1°C, as these temperatures are best to extract as much soluble protein as possible from the muscle meat.

VACUUM PACKAGING MACHINE

For vacuum packaging the meat product has to be placed into a vacuum bag. Air is removed from the bag by means of the vacuum packaging machine and the bag then sealed. Special vacuum packaging machines can operate with so called gas-flushing, where a mixture of gas is injected after evacuating the air. Such protective gas atmospheres inside the product package inhibit bacterial growth and stabilize the meat colour. The gas mixtures usually contain CO_2 and N_2.

Mixer / blender

Mixers are used to blend meat and spices, or coarse and finely chopped meat. The machine generally consists of a rectangular or round bottom vessel through which two parallel shafts operate. Various paddles are mounted on those shafts to mix the meat. The mixer is discharged through tilting by 90 degrees. Some mixers are designed as vacuum mixers, as the mixing under vacuum (exclusion of oxygen) has advantages for the development of desirable product colour and texture.

Emulsifying Machine (colloid mill)

The emulsifier serves for the preparation of very fine meat emulsions. Its functional parts are a perforated plate, attached to which two edged blades are rotating (rotor blade). Next to the blades there is a centrifugal pump that forces the pre-ground meat through the perforated plate. Most emulsifiers are vertical units. Compared to the bowl cutter the emulsifier operates at much higher speed, producing a finer emulsion-like mix. The emulsifier is also perfectly suited to produce semi-processed products such as pig skin emulsions.

In these machines ice flakes are continuously produced from potable water. Ice is needed in meat processing for some types of meat products. Water, added in the form of ice, is an important ingredient in order to enhance protein solution and to keep the temperature of the meat batter low. Ice flakers with in-built UV-water-disinfection device are available for areas with unsafe water supply.

Frozen Meat Cutter

The purpose of cutting frozen meat blocks into smaller pieces is to make frozen meat suitable for immediate comminuting in grinders, bowl cutters etc. without previous thawing. There are two types of machines for the cutting of frozen meat blocks, working either with knives cutting in vertical direction (guillotine principle) or using rotating drums with attached sharp knives. In the guillotine-type machines a knife head is driven hydraulically and even the hardest frozen products can be cut into small pieces, either meat cubes or meat strips. Rotary frozen meat cutters operate according to the principle of carving out particles from the frozen meat blocks. The rotary drums can be equipped with knives capable of cutting out pieces of frozen meat from large fist-size to small chip-size.

CURING OF ENTIRE MEAT PIECES

Besides the curing of chopped meat mixtures, entire pieces of muscle meat can be cured. However, due to size the curing substances cannot instantly react with the meat pigments as is the case in chopped meat mixes. Hence various curing techniques are applied. The final products of curing entire meat pieces are either cured raw fermented products or cured cooked products. The curing system to be used depends on the nature of the final product (uncooked or cooked). There are two systems for curing entire meat pieces, dry curing and wet curing ("pickling") and the type of the final product determines which system will used. In dry curing a curing mix is prepared containing salt or nitrite curing salt, together with spices and other additives. The pieces of meat are rubbed with this curing mix and packed in tanks. The curing mix gradually permeates into the meat, which can be a lengthy process ranging from several days to several weeks.

Dry curing is exclusively used for the fabrication of cured raw fermented products, in particular those with a long ripening period.

The second method of curing meat pieces is wet curing, also called pickling, which involves the application of curing brine to the meat. For the manufacture of the brine, curing salt and spices, and other additives if required are dissolved in water. The meat cuts are packed in tanks and brine is added until all pieces are completely covered. A temperature of +8 to +10°C for the curing room is recommended as lower temperatures may retard curing. For equal penetration of the brine, the meat is cured for periods ranging from several days to two weeks depending on the size of the cuts and curing conditions. After completion of the curing, ripening periods for the products follow for taste and flavour build-up.

Wet curing by immersion of meat pieces in brine is primarily used for the fabrication of cured raw fermented products with shorter ripening periods.

An alternative and quick way of wet curing is to accelerate the diffusion of the curing substances by pumping brine into the meat tissue ("injection curing"). For this purpose brine injectors with perforated hollow needles are used. The injection of brine into the muscles can be done manually by using simple pumping devices. At the industrial level semi-automatic multi-needle brine injectors are used which achieve very even distribution of the curing ingredients and can reduce the curing period (equal distribution of the curing substances or "'resting period") to less than 48 hours.

In addition, most injection cured meat pieces which are to be processed into cured-cooked products (such as cooked hams etc), are submitted to a tumbling process. Tumbling further accelerates the brine penetration throughout the meat prices and "resting periods" are not necessary.

Smoking

Smoke for treatment of meat products is produced from raw wood. Smoke is generated through the thermal destruction of the wood components lignin and cellulose. The thermal destruction sets free more than 1000 desirable or undesirable firm, liquid or gaseous components of wood.

These useful components contribute to the development of the following desirable effects on processed meat products:

- Meat preservation through aldehydes, phenols and acids
- (anti-microbial effect)
- Antioxidant impact through phenols and aldehydes
- (retarding fat oxidation)
- Smoke flavour through phenols, carbonyls and others
- (smoking taste)
- Smoke colour formation through carbonyls and aldehydes
- (attractive colour)
- Surface hardening of sausages/casings through aldehydes (in particular for more rigid structure of the casing)

The most known undesirable effect of smoking is the risk of residues of benzopyrene in smoked products which can be carcinogenic if the intake is in high doses over long periods. With normal eating habits, a carcinogenic risk is normally not associated with moderately smoked food such as smoked meat products.

Depending on the product, smoke is applied at different temperatures. There are two principal smoking techniques:

- Cold smoking
- Hot smoking

The principle of both methods is that the smoke infiltrates the outside layers of the product in order to develop flavour, colour and a certain preservation effect.

Cold Smoking – This is the traditional way of smoking of meat products and was primarily used for meat preservation. Nowadays it serves more for flavour and colour formation, for example in sausages made from precooked materials such as liver sausage and blood sausage.

The combination of cold smoking and drying/ripening can be applied to fermented sausages and salted or cured entire meat pieces, in particular many raw ham products. In long-term ripened and dried hams, apart from providing colour and favour, the cold smoking has an important preservative effect as it prevents the growth of moulds on the meat surfaces.

The optimal temperature in "cold" smoking is 15 to 18°C (up to 26°C). Sawdust should be burned slowly with light smoke only and the meat hung not too close to the source of the smoke. Cold smoking is a long process which may take several days. It is not applied continuously, but in intervals of a few hours per day.

Hot Smoking – Hot smoking is carried out at temperatures of +60 to 80°C. The thermal destruction of the wood used for the smoking is normally not sufficient to produce these temperatures in the smoking chamber. Hence, additional heat has to be applied in the smoking chamber.

The relatively high temperatures in hot smoking assure a rapid colour and flavour development. The treatment period is kept relatively short in order to avoid excessive impact of the smoke (too strong smoke colour and flavour).

Hot smoking periods vary from not much longer than 10 minutes for sausages with a thin calibre such as frankfurters to up to one hour for sausages with a thick calibre such as bologna and ham sausage and products like bacon and cooked hams.

Products and smoking – Cold smoking is used for fermented meat products (raw-cured ham, raw-fermented sausage) and precooked-cooked sausage (liver and blood sausages). Hot smoking is used for a range of raw-cooked sausages, bacon and cooked ham products. Smoke treatment can only be applied, if meat the products are filled in casings permeable to smoke. All natural casings are smoke permeable, as are cellulose or collagen basis synthetic casings.

Smoke permeable casings can also be treated using a new technology, where a liquid smoke solution is applied on the surface. This can be done by dipping in solution, showering (outside chamber) or atomization (spraying inside chamber). Polyamide or polyester based synthetic casings are not permeable to smoke. If smoke flavour is wanted for products in such casings, small quantities of suitable smoke flavour (dry or liquid) are added directly to the product mix during manufacture.

PLAIN MEAT PROCESSING UNDER BASIC CONDITIONS

Meat processing can be carried out even under basic conditions at a very small-scale level such as a village or rural setting. Such small-scale meat products manufacture can be recommended from the nutritional point of view, as it enables communities to fully utilise animal food materials derived from small-scale slaughtering and allows for extension of meat products with locally available suitable plant food. This type of meat processing also provides the option to manufacture products with increased shelf life, *e.g.*, through drying or other preserving measures, *e.g.*, salting, smoking, fermentation or heat treatment enabling communities to cater for periods meat might not be available.

Basic processing equipment can be manually operated for the smallest units with little throughput. The availability of single-phase electrical power allows for larger quantities to be processed using basic electrically powered equipment. In least-developed and developing countries the high capital investment required to construct a purpose-built facility may not be feasible. In these cases the conversion of a suitable building of good standard can be acceptable as long as basic hygienic measures are introduced and followed.

Facilities

A purpose built area is highly recommended for commercial small-scale meat processing. As a guide, facilities should have the following minimum characteristics:

- The floor must be of hard, smooth and waterproof material in anti-slip finishing. Materials such as high-grade concrete, terrazzo (= concrete mixed with small stones and smoothened on the surface after hardening), floor tiles or plastic coating are recommended. Floors must be sloped to drain off surface water.
- To avoid difficult-to-clean spots, rounded junctions of walls and floors are strongly recommended.
- The surface of interior walls must be waterproof and smooth and should be covered by wall tiles or coated with washable paint up to a height of 2.0 m.
- The surface of the ceiling should be smooth, flat, sufficiently high and preferably also painted with washable paint.
- To avoid high humidity in the facility, sufficient ventilation must be guaranteed.
- All windows, doors and other ventilation openings need to be insect proof.
- A drainage system should be built-in to allow rapid disposal of effluents. The openings inside and outside must be covered with a removable netting to allow collection of solid particles and prevent rodents and other small animals from entering the system. If facilities are not in use, all inlets and outlets need to be closed hermetically.
- Areas of clean operations (meat cutting and processing) and dirty operations (cleaning of intestines) must be divided by a wall or carried out in separate rooms. In warmer climates the dirty area can be semi-open, in colder regions a small room is needed.
- Potable water must be available for cooking of raw materials and processed products and cleaning of facilities and equipment.
- Wastage such as solid particles from the drainage system and the

content of the gastro-intestinal tract could be collected and converted into biogas to be used for heating and light.

Equipment and Tools

Without electrical power the meat processor is limited to manually operated equipment. A meat grinder is needed to cut meat and non-meat ingredients to the desired size. A manual sausage stuffer allows for easy filling of sausage mixes into the casings. Other tools like knives, hand bone saw, sharpening steel, buckets, one big and several small containers (plastic) and meat hooks are also needed. For grating of peeled roots, stainless steel grating plates are very helpful tools as they are unbreakable.

Cleaning and Hygiene

Even when processing is done under such simple and basic conditions, due attention needs to be paid to hygiene. Cleanliness of facilities, installations and personnel is an essential pre-condition in all food processing. A sufficient amount of potable water, boiling facilities (charcoal cookers, gas stoves) and detergents and disinfectants must be available. Raw materials brought to the processing area must be fresh, clean and disease-free. It is essential that the facilities, machinery and tools are properly cleaned before and immediately after processing. A high level of personal hygiene is required and therefore essential that staff involved in processing must be in good health and undergoes regular medical check ups. Staff members should never enter the food processing area in street clothes and shoes, but wear clean protective clothing, and also carefully attend to personal hygiene.

Product Range

The range of products depends on the climatic conditions under which the processing is done, when manually operated equipment is used and refrigeration facilities are lacking. In tropical and subtropical countries with temperatures of above +25°C during the day and no significant decrease in temperature during the night the production is limited to products which are immediately consumed or products which are dried, salted and smoked in simple smoking units. Fresh coarse sausages, burgers and meat-mixes (meats/vegetables) are typical products for immediate consumption. Dried and/or fermented sausages of small diameter for rapid drying and dried meat sticks can also be produced.

In moderate climates, smaller raw-fermented hams and larger calibre dried-fermented sausages can be produced during colder seasons. Some of the precooked-cooked products, such as coarse liver sausage, stuffed in natural casings of smaller calibre can be produced for immediate consumption.

PROCESSED PRODUCTS MADE OF CHICKEN MEAT

Processed chicken is chicken meat that has either been mechanically recovered from a chicken carcass or is made of a combination of chicken meat and skin. Many different types of processing can be done on other meats, but processed chicken is primarily either mechanically recovered or is made from ground up chicken meat and formed into shapes. Although they conjure up images of machinery and factory lines, processes can still be completed by human workers. Any meat that has gone through a process such as drying, fermenting, curing, or pre-cooking can be classified as processed meat.

The most common bits of a chicken used for food are the breasts, legs, and wings, but processing makes other parts of the chicken edible. During the processing of a full chicken carcass, the valuable cuts such as the breasts, legs and wings will be removed by a trained employee working with a knife, and then the leftover carcass is often used to mechanically recover the leftover meat.

Depending on how the more valuable cuts of meat are going to be used, they may or may not be processed into another form. Foods such as chicken frankfurter are made from processed chicken, using a mixture of lean meat and chicken skin. Chicken bologna is also made from these processed parts.

Mechanical recovery is one way of making processed chicken that has attracted media attention. This method takes the chicken carcass leftover after the removal of the most valuable cuts, and then runs it through a machine. The carcass is ground up, along with cartilage, bones, and skin, in order to recover the meat that can be still garnered from the chicken. The processed chicken then comes out in a pink mousse, brimming with bacteria, waiting to be treated with ammonia to kill the microorganisms. The chicken is then re-coloured in order to make it conform to expectations of what chicken looks like and re-flavoured to remove the ammonia taste.

This mixture of processed chicken is then used in a variety of different food products in place of more expensive lean chicken. Products which make use of mechanically processed chicken include chicken ham in the US, or chicken meat balls in Asia. This process is often thought to be used in the making of chicken nuggets, but while the meat is still technically processed in most chicken nuggets, it cannot be said to be mechanically processed. Other types of processing include drying, frementing, curing, or pre-cooking.

Chicken Sausages

A sausage is a food usually made from ground meat with a skin around it. Typically, a sausage is formed in a casing traditionally made from intestine, but sometimes synthetic. Some sausages arecooked during processing and the casing may be removed after. Sausage making is a traditional food preservation technique. Sausages may be preserved by curing, drying, or smoking.

Sausages Containing Meat Mixes Including Chicken Meat

Chicken meat is often used to partly substitute the more expensive red meats in meat products especially of the raw-cooked type such as luncheon meat, bologna or hotdogs. In such cases the chicken meat percentage can vary substantially. As a good manufacturing practice, the percentage of chicken meat should be indicated for consumer information, as such products are normally perceived as pork or beef products, not containing chicken meat. Also some liver sausages can contain larger quantities of chicken meat. These products are commonly labelled as "Chicken Liver Sausage", although in many cases the liver and animal fats derive from pork.

Sausages and Other Products with 100 per cent Chicken Meat

Besides the use of chicken meat in mixed red meat products, there are many well established and popular products which contain chicken meat only. When processed chicken and turkey meat products were introduced on a broader scale a few decades ago, traditional red meat recipes where simply modified and red muscle meat was replaced by poultry meat and pork fat by fat rich chicken skin. For these poultry products such as chicken frankfurter, chicken bologna, etc., non-meat ingredients and the processing technologies remain basically the same as for the corresponding processed red meat products. Manufacturers even endeavour to make chicken and turkey sausages similar to red meat sausages in taste and flavour, but point out the health benefits of poultry products.

Chicken frankfurter and chicken bologna are finely comminuted products, which can be considered as raw-cooked products. Lean chicken meat provides the proteins and chicken skin replaces the fat to be finely dispersed in the sausage batter. Filled in small casings (18-22mm), this typical raw-cooked sausage mix is the basis for chicken frankfurters, when filled in larger casings (40-60mm), for chicken bologna. The mix also serves as the basis for products where coarse chicken meat (either diced or ground) is blended with batter and filled in casings of 60-80mm or cans. These products may be named chicken or turkey ham sausage, chicken or turkey roll, etc. Chicken meat balls, a product in high demand in the Asian region, are also of the same category. They are manufactured based on the method used for traditional meat balls from red meat. Another chicken meat product, which resembles the cooked hams made from pork in both manufacture and appearance is called chicken ham (raw meat material may come from all parts of the chicken carcass) or chicken breast (in this case only chicken breast parts should be used). The meat material is tumbled together with brine containing curing salt, phosphates and spices, and either pasteurised when filled in casings or moulds or sterilised when filled in cans.

OTHER CHICKEN MEAT PRODUCTS

Coated/ Breaded Products

In addition to chicken sausages, the chicken meat industry also developed new products, which contributed significantly to meeting the global increase in demand for poultry meat. These can be compared to a few examples from the red meat and fish sector, *e.g.*, breaded and fried meat slices of pork or mutton known as "Escalope" or "Wiener Schnitzel" and in the fish sector as "fish fingers". The characteristic of such products is the coating of meat surfaces with flour, fat/flour mixes and/or breadcrumbs, etc. In the poultry sector, similar products include chicken nuggets (ground meat mix), chicken sticks or fingers (muscle strips) or chicken schnitzel (breast muscle slices).

After the meat or meat mix is portioned, each portion is pre-dusted by applying a thin layer of dry flour on the meat surface. This serves to firmly absorb the batter and the breading in the following steps of the processing. Battering consist of dipping the meat pieces in a semi-liquid mixture of oil, eggs, water and spices. Breading is the coating with flour, fat/flour mixes and/or breadcrumbs. The final step in this process is heat treatment to stabilise the coatings on the meat surface. This short heat treatment in hot fat/oil (approx. +175°C) as part of the processing must be seen as a pre-treatment only and does not cook the product. The final heat treatment is carried out by the consumer right before eating.

PRINCIPLES OF MANUFACTURE

The manufacture of raw-fermented sausages at the small to medium scale meat industry level is outlined hereunder. These sectors often lack a full range of comminuting equipment and in particular equipment for accurate climatisation during fermentation and ripening and therefore face more technological challenges than larger, well equipped industries.

Raw Materials

The processing of raw-fermented sausages is dominated by biological and biochemical processes and raw meat materials of excellent hygienic quality are a precondition for the correct functioning of such processes. Lean meat from a variety of animal sources such as cattle, pigs, horses, donkeys, camels, sheep or goats can be used. The lean meat can be from older adult animals, as water content and water holding capacity of such meat is lower, which supports the necessary drying processes during fermentation and ripening. All meat used must be chilled for some time to reach its lowest ph-values. Beef meat should have pH-values at 5.4-5.5, pork meat 5.7-5.8. All lean meats for raw-fermented sausages need extra careful trimming of sinews and softer inter-muscular fatty tissue. Remaining sinews will remain tough and are not desired by consumers.

In most products fresh chilled pork backfat is used as it is firm and dry and remains stable without pronounced rancidity even after prolonged ripening periods. Softer inter-muscular fatty tissue should not be used as it cannot be chopped to clearly defined particles and would result in somewhat blurred unclear appearance of slices of the final products. Soft fat also increases the risk of early rancidity. If fats from other species of slaughter animals are used, only firm body fats should be considered.

Importance of Bacteria

Bacterial starter cultures have a variety of functions including:

- Boosting acidity (decreasing pH)
- Intensify the curing colour (acid environment catalyses curing reaction)
- Counteract rancidity of fats (due to enzymatic impacts)
- Development of flavour and taste
- Texture improvement of ripened products (by supporting formation of protein gel in sausage mixes).

Over the years, mainly bacteria belonging to the groups of Lactobacillus, Pediococcus, Staphylococcus and Streptococcus have been identified and cultivated for commercial starter cultures, as they proved to provide the best results in terms of producing lactic acid, developing ripening flavour, and are generally harmless in terms of product spoilage and impact on consumers' health. Depending on the desired taste, texture and appearance of the product, specific cultures are selected. The use of Lactobacillus results in fast acidification to lower pH-values, the use of Pediococcus leads to slower and milder acidification. Selected Staphylococcus strains cause a speedy reduction of nitrite, stable curing colour and reduced risk of fat rancidity, especially in products fabricated with Glucono-delta-Lacton.

In most cases mixtures from different strains are used in order to achieve the best product specific results, for example in sausages with normal diameters (35-70 mm) an even mixture of Lactobacillus and Staphylococcus can be used to achieve the product-typical flavour, texture and taste. In sausages with of larger diameter (70-100 mm), the starter culture mixture normally contains a lower amount of Lactobacillus and a higher portion of Staphylococcus, as these products need more time to reach microbial growth inhibiting moisture contents. The strong potential of Staphylococcus to stabilise curing colour and fats is helpful in this context.

Importance of Salt, Curing Agents and Sugars

One of the main targets during fermentation and ripening of raw-fermented sausages is the reduction of their water content. The moisture to be reduced is exclusively from the muscle meat which has a water content of around 80 per cent. The addition of salt lowers the a_w value of the mix by absorbing water, which presents an initial hurdle for unwanted bacteria. Furthermore, in the presence of salt, salt-soluble proteins are extracted from the small lean meat particles after grinding and chopping. These solubilised or gelatinous proteins act like an adhesive between the interfaces of lean meat and fat particles in the meat mix. The result is an increasingly firm structure with progressive ripening and drying of the products. The average quantity of salt added to raw-fermented sausages should be between 26-30 g/kg (2.6-3.0 per cent) but not below 26 g/kg (2.6 per cent). It should be noted that the salt content in percent in the final products will always be higher than in the initial mix, as these products lose a substantial amount of water. Salt contents in final products can be from 3 - 4.5 per cent depending on the initial salting.

In raw-fermented sausages, salt is also used as a carrier for the curing agent, normally sodium nitrite. This curing agent is not only responsible for the development of a typical red cured meat colour, but also has bacterial growth inhibiting properties, especially on some pathogenic bacteria. In raw-fermented sausages with a slow decrease of pH-values and prolonged ripening periods, nitrate can also be used as a curing substance. The use of both, nitrite and nitrate results in similar colour and taste. The main difference is that nitrate must first be reduced to nitrite by bacteria, which is a time-consuming process and hence only applicable to long-term ripened products. The slowly progressing acidity in such sausages allows the bacterial breakdown of nitrate to nitrite. The following reduction of nitrite to nitrogen oxide (NO), which is the substance effective in the curing reaction, is a relatively fast chemical process. The use of nitrate, mixed with nitrite is favoured by some processors as it is associated with better colour and flavour.

From the technical point of view, the purpose of adding sugars is to facilitate and strengthen the fermentation by bacteria. Provision of a sweet flavour to counteract acidity in the final product is normally not intended. The bacterial breakdown of sugars results in the accumulation of lactic acid and in a low pH-value (acidification) as well as the development of a typical flavour. In order to support this process, lactic acid producing bacteria can be added to the sausage mix. Simple sugars such as dextrose or fructose support an early drop in pH-values as they are easily broken down by bacterial action.

The breakdown of lactose is slower and takes longer. Often a mixture of different sugars is used. Another sugar-based additive is GdL (Glucono-delta-Lactone), which accelerates and intensifies the acidification process by reacting to glucono-acid in the presence of water (muscle tissue water). It is preferably used in semi-dry and/or spreadable products, which are not for long-term ripening and storage, but for consumption within a short period after production.

Production Methods

As a rule of the thumb, raw-fermented sausages are fabricated with 20-35 per cent fatty tissue and 65-80 per cent lean meat, from one or more than one animal species, *e.g.*, beef and pork or pork only or beef only. Other variations are also possible. If fatty tissue other than pork back fat is used the percentages for the fat are usually lower. The techniques of comminuting of meat and fat for raw-fermented sausages differ from other meat products.

Raw-fermented sausages may be composed of coarse, medium or tiny meat and fat particles. The degree of chopping can be visualised by the size of the fat particles in the final product.

Some traditional Mediterranean (Italian, Spanish, French, etc.) salamis are chopped coarsely (6-12mm), but the majority of raw-fermented sausages are chopped moderately (2-5mm). Only a few semi-dry and/or spreadable products are finely chopped.

In small to medium-sized processing, there are two methods of manufacture of raw-fermented sausage mixes, which basically differ by the method of comminution of the raw materials. Applying a simple comminuting method, only meat grinders are used to prepare the sausage mixes. In more advanced techniques meat grinders and bowl cutters are used.

- *Method 1:* In small-scale operations with only meat grinding equipment available, production is restricted to ground sausage mixes. The lean meat needs to be thoroughly chilled (+1°C) or even slightly frozen. The fat portion should be cut into small and uniform dices (10-20 mm, domino chip size) and frozen (-12°C) in order to obtain clearly and evenly cut particles in the initial chopping of the sausage mix. Clearly cut particles of firm solid fat also avoid greasing of the casing from inside, which would make drying more difficult. Firstly, part of the lean meat is minced 3-5mm (approx. 30 per cent) and the remaining lean meat is cut into small pieces (20-50 mm). The chilled meat pieces and frozen fat dices are thoroughly mixed with all additives (curing salt, sugars, starter cultures, spices, etc), before the minced meat portion is added and incorporated in the mixture. The entire mixture is now passed through the meat grinder (disc size 3-6 mm), packed into the sausage stuffer and stuffed into casings. Delays leading to warming up of the mixture need to be avoided as this would result in greasing during the stuffing.

For the stuffing, natural or artificial casings can be used. Typical natural casings, depending on the desired sausage diameter, are those derived from the small intestines of pigs, sheep, cattle or horses. Artificial casings used are fibrous or collagen casings. One important requirement for casings used for raw-fermented sausages is to closely adhere to the sausage mix not only after filling but also during the drying period when sausages shrink. The casings used must be water vapour permeable, otherwise no drying during fermentation and ripening can take place and the products would spoil. The required conditions are met by natural casings, and fibrous and collagen casings.

- *Method 2:* With a bowl cutter available, a different technology can be applied. With this method 50 per cent of the lean meat material is minced (3 mm) and kept at 1°C. The remaining 50 per cent of the lean meat is cut into pieces of 30-50 mm diameter and slightly frozen (-10°C). As per method 1, the fat is cut into small dices (preferably 10-20 mm, domino chip size) and also frozen (-12°C). Firstly, the large pieces of frozen lean meat are chopped. If starter cultures are used, they must be added at this stage. After several rounds of the frozen lean meat in the bowl cutter, the frozen fat is added together with the spices and sugars and chopping is continued at a medium speed until the fat has reached the desired particle size. Then the minced chilled meat is added under low chopper speed until an even distribution is achieved. In the next step, the nitrite curing salt is added and mixed at low speed for at least 6-8 rounds until a final temperature of around -5°C is reached. This mix temperature should not be exceeded in order to avoid the greasing of the interior of the filling funnel and casings.

When lean beef and pork is used for the above raw-fermented sausage fabrication, the beef should be chosen for the 50 per cent lean meat portion to be minced, while the pork portion is preferably used frozen.

FRESH PROCESSED MEAT PRODUCTS

This group comprises meat mixes composed of finely comminuted, minced or sliced muscle meat, with varying quantities of animal fat adhering to the muscle meat or added separately. Flavouring is done by adding common salt and spices; curing is not practiced. In many products other non-meat ingredients are added in smaller quantities for improvement of flavour and binding, in low-cost versions larger quantities are added to extend the existing volume. The characteristic of this group is that all meat and non-meat ingredients are added fresh (raw), either refrigerated or non-refrigerated. The heat treatment (frying, cooking) is only applied immediately prior to consumption to make the products palatable. In many instances, the consumer cooks the products prior to serving and products are consumed hot. Most of the fresh meat mixes are filled in casings, which defines such products as sausages. If other portioning is customary, the products are known as burgers, patties, kebab, etc.

Patties are formed from minced meat usually in a disc-like shape with diameters of 80-150mm and 5-20mm height. In commercial fast-food outlets the common name ishamburgers or simply burgers. Originally, burgers were made from *beef* (preferably lean cow meat), but in recent years *chicken* and *mutton* burgers have become more common. Other animal tissues such as fats or connective tissue/tendons can also be part of the mixture, with quantities depending on the type and quality of the products. In industrial manufacture, these tissues could have been previously separated from the lean meat and are added again in defined quantities to ensure identical chemical composition (protein, fat, water) of all products. A common feature of burgers is that during mincing (1-3mm disc) and consecutive blending, salt and spices (mainly black and white pepper, in some instances also herbs, garlic or onions) are added. In some cheaper industrial formulations textured soy protein is commonly used as a non-meat ingredient in quantities up to 25%. Other non-meat ingredients suitable for this purpose could include rusk, breadcrumbs and dried flakes from roots and tubers.

Burgers are stored frozen and individually pan-fried before consumption. Ideally, internal temperatures of 80°C should be reached to destroy food poisoning agents potentially present in the raw meat mixes. Burgers are often served on bread rolls or buns with slices of cheese, mayonnaise, mustard, green salad, etc.

The Kebab is a Middle East product, but popular in many places and usually eaten in pieces of flat white bread with yogurt sauce or sheep cheese. These preparations of kebab are also known by the name of doener or gyros. The term "kebab" refers to processed meat on skewers. Kebabs are usually made of sliced lean meat from veal, mutton or chicken or mixes of them. The lean meat has been marinated (mixture of salt, spices and oil) and the marinated meat pieces are arranged around a skewer bar. The usual quantity of meat on the skewer is 3-4 kg.

For preparing the product for consumption, the skewer is slowly rotated in a vertical position close to a source of heat. Traditionally glowing charcoal was positioned on the backside of the skewer in a metal basket. Nowadays gas elements, electro coils or infrared devices are used. The outside layers of the meat bulk, once they are sufficiently heated (slightly crispy), are carefully trimmed off as thin slices.

In doing so, the deeper layers, which are still uncooked, will be exposed to the heat and trimmed off when cooked. The process is repeated until all meat has been trimmed off. A special kebab is produced using minced or finely comminuted meat mixes similar to patty mixes. This type of kebab must be heat treated (coagulated) prior to final roasting to make sure that the big chunk of meat firmly sticks to the vertical skewer and maintains its shape and position.

Other varieties of kebabs are prepared in individual portions in fast-food outlets. These kebab types usually consist of fresh or marinated small meat dices or flakes on a skewer. Some variations can contain visible portions of vegetables (bell pepper, onions, etc) or even liver/kidney pieces. A typical marinated meat-only variety is the Greek souflaki containing veal or lamb meat which is marinated with lemon juice, herbs and garlic. Souflaki is grilled over charcoal. Another variety, where often vegetables and liver/kidney pieces are included, is known asshashlik. This type is briefly fried (browned) in little oil and simmered in a heavy sauce. These individually portioned kebab varieties are nowadays also available raw (fresh or frozen) as convenience products and prepared by customers at home.

Fresh Sausages

Fresh sausages probably represent the oldest form of processed meat products. Their production could be carried out everywhere where animals were slaughtered, which produced both the meat and the casings. In the simplest way of manufacture, no tools other than knives are needed. Fresh meat and fat are mixed with salt and spices and stuffed into natural casings derived from small intestines of slaughter animals. Higher quality fresh sausages are primarily composed of lean meat and fat. In some low-cost formulations non-meat extenders are also used.

Fresh sausages products are well suited for small-scale meat processing outlets, as all ingredients including casings can be generated or procured locally. The manufacture can take place with basic meat processing tools and machinery. These sausages do not undergo heat treatment at processor level, but are roasted, fried, boiled or otherwise heat treated before consumption upon demand by consumers or by consumers themselves.

Meat and Non-meat Ingredients

The animal tissues (meat and fat) used in fresh sausages can originate from different animal species (pigs, cattle, small ruminants, game, poultry, fish). The meat selection and lean/fat ratio vary, depending on cultural preferences and consumer expectations. Most fresh sausages are coarsely chopped products. Hence the lean meat should be free of tendons or hard connective tissue and only solid fats (beef body fat, pork back fat) should be used. The hard connective tissue would remain relatively tough in the ready-to-eat product and soft fatty tissues would make the product greasy. In addition, the fat content in the final product should not exceed 25%, as otherwise the shrinkage by melting fat during frying or cooking would be high.

In traditional recipes only common salt is used (10-15 g per kg raw material) as red cured meat colour is not required in these products. Hence curing salt is unnecessary. The most common spices used in fresh sausage production are pepper, mace, coriander, red chilli, cardamom, ginger and cumin. Depending on availability and desired flavour and taste smaller quantities of onions and/or garlic can also be added. Sausages composed primarily of meat and fat are "frying sausages", which are popular around the globe. Those made from beef or pork or containing mixtures of both are the best known.

PROCESSING OF HIGHER QUALITY FRESH SAUSAGES

Raw fresh lean meat and fatty tissue are the main components of fresh sausages. Typical examples for this sausage type exist in all regions of the world. The most popular products are:

- *"bratwurst"* which means "frying sausage" in Central Europe
- *"longaniza"* and *"chorizo criollo"* in countries with Spanish tradition
- *"merguez"* in Northern Africa and Middle East
- *"breakfast sausage"* in countries with British tradition
- *"boerwors"* in South Africa

For the manufacture of coarsely chopped fresh sausages lean meat and fats are cut by hand into pieces, mixed with salt, spices and other non-meat ingredients (step 2) and minced in a meat grinder (step 3), using a grinder disc with the desired size of disc perforations (4 to 6 mm).

Other types of fresh sausages are composed of finely chopped raw materials or a combination of coarse meat and finely chopped portions. In these variations additional ingredients such as eggs, milk, starches, etc. can be used, primarily to improve the binding of the final product. For the preparation of such finely chopped meat mixtures a bowl cutter is necessary.

The use of a bowl cutter also enables the incorporation of larger quantities of extender materials for low-cost recipes. After grinding, the mixture is usually stuffed into thin or medium size calibre natural casings of the "edible" type. These casings, derived from the small intestines of pigs or sheep, are either freshly prepared from local slaughter, or salted and stored until used. In any case, these fresh natural casings need to be rinsed with sufficient quantity of clean water before being used for stuffing. The casings are filled almost to their maximum capacity (step 5) and thereafter divided into shorter units of the desired size by linking and twisting (step 6).

Natural casings can also be replaced by edible collagen casings of similar diameter. This allows for better standardisation of sausages and larger volumes of production. In the absence of casings the mixture can also be shaped into meat rolls (also known as skinless sausages), meat balls or burger patties. This is done either by hand or by using simple tools.

Storage and Preparation for Consumption

Fresh sausages are highly perishable products and subject to fast microbial spoilage and oxidative rancidity. They should be heat-treated and consumed as soon as possible after production, or must be stored immediately under refrigeration. Their maximum storage life is normally three days at +4°C or below. If the product is deep-frozen at -18°C, the storage life can be extended up to three months. However, one difficulty associated with frozen storage is the oxidative rancidity. Storage in vacuum bags can prevent the fast onset of rancidity.

Processing of Local Low-cost Fresh Sausages

Low-cost variations of fresh sausages are widely available. In these variations, available vegetables and other fresh plant ingredients and their derivates (bell pepper, bread crumbs, soy concentrates/textured vegetable protein, potato, fresh cassava, dried cassava flakes, rusk, etc) are used as extenders. Small amounts of binders such as starches and flours are also common in some of the products.

In cases where smaller or larger quantities of fresh vegetables and other fresh plant ingredients are incorporated, special pre-treatment of these non-meat materials must be carried out. Onions and garlic must be thoroughly peeled; bell pepper must be washed and all seeds and stem parts must be removed; fresh cassava and potatoes must be washed and peeled. All components are cut into uniform small pieces and mixed with meat and spices. For a good degree of blending of such fresh plant/vegetable extenders the mixes containing all ingredients are minced 3-5 mm before stuffing them into available casings.

4

Basic Technology of Meat Comminution

Comminution is a procedure which modifies the properties of fresh meat, so that the finished product consists of small meat and fatty pieces, or it can be finely subdivided to obtain a meat batter. This meat batter is a viscous mass, with many properties of an emulsion. A major problem in manufacturing emulsion-type sausages is the tendency of fat to be separated during heat treatment. The sausage batter, at least in part, is formed of a "fat/water emulsion" (meat emulsion or, better to say, dispersion), employing salt-soluble proteins of meat as emulsifying agents.

Comminution, the basic method for sausage making can be: coarse, to produce non-emulsified sausages like salamis and summer sausages, or fine, to produce emulsified sausages like frankfurters, bologna, etc..

Coarse comminution is used to manufacture sausages with a coarse texture with small pieces of meat and fatty tissues visible. For Meat I and for Meat II a 15- to 25-mm grinder plate is used while trimmings and fatty tissues are passed through a 2- to 6-mm plate so that the connective tissues are less visible and more digestible. Meat I and Meat II are passed through the grinder first, followed by the trimmings and finally the fatty tissues. This avoids fat separated from the fatty tissues greasing the surface of meat pieces thereby decreasing their binding ability. The meat should be well refrigerated (2–3°C) and firm, so that it will be cut cleanly by the grinder knives and not squeezed.

Correctly pre-ground meat is transferred into the mixer or mixed manually for formulation. Nitrite salt and phosphate are added and the mixing starts. All other non-meat components are added during mixing, finishing with ascorbic acid. When the mixture is uniform the sausage mixture is formed.

Fine comminution used to manufacture emulsion-type sausages is performed in a high-speed cutter. The knives must be kept thoroughly sharpened and properly adjusted to the bowl. During comminution bundles of fibres, myofibrils and filaments are separated and disrupted and the size of meat and fatty tissues is continuously reduced. The muscle tissue, having a stronger structure than fatty tissue, determines the total comminution time required. Heat produced during comminution helps the salt-soluble proteins to concentrate at the fatglobule surface and come in contact with water from the protein matrix. Higher speeds produce a better distribution of fat due to the temperature rise but excessive temperatures (more than 15°C in pork and 18°C in beef) can partially destroy the protein membrane which surrounds fat particles, and cause a breakdown of the sausage mixture. Unprotected melted fat migrates to the surface, forming easily visible fat pockets between the sausage surface and the casings. The melting point of fat is the parameter for determining the maximum temperature for the meat batter. Overcomminution can also provoke breakdown. As the fat particle size continuously decreases, there is a proportional increase in the total surface of fat particles requiring proteins. Generally, as particle size decreases more meat and meat-extender proteins are required to form a heat-stable emulsion.

Frozen and chilled meat for different products. Frozen meat must be sliced in a frozen-meat slicer prior to grinding. The block of meat is placed upon the carriage and passed under a large knife which cuts the meat into large slices or smaller meat flakes.

Frozen meat is not suitable for coarse comminuted sausages but may be used in manufacturing fine comminuted meat products. So that the optimal temperature (3–5°C) for extraction of salt-soluble proteins is reached, only water without ice is added in the cutter.

Some meat extender must be added to bind juices from the frozen meat which will be separated during heat treatment.

Chilled meat is more suitable for both coarse and fine sausages as the meat batter can be more readily maintained at the optimal temperature. Another advantage is that neither a frozen-meat storage facility (minimum -18°C) nor a frozen-meat slicer is necessary.

Methods of addition of salt, additives and spices. The aim of using salt and/or nitrite salt in manufacturing comminuted meat is to separate the maximum quantity of salt-soluble proteins from the muscle cells. Optimum salt concentration for total extraction is 5–6 percent, and the optimum temperature is 3–5°C. Salt must be sprinkled on the pre-ground meat in the bowl as soon as possible. By mixing meat and salt at the slowest speed, salt starts to extract proteins from broken muscle cells. Half the ice water (1:1) is added to speed this process and control the temperature rise.

Additives contribute to improving and intensifying some properties of meat proteins, especially water-holding, water-binding and emulsifying capacities. In order to achieve this they are sprinkled on the meat after the salt has completed its effects, but always before adding fatty tissues.

To achieve uniformly and well-flavoured comminuted meat products, spices must be properly sprinkled in the meat batter and are added after additives, but always before fatty tissue to avoid them sticking to it causing a non-uniform colour. All spices for sausage formulation should be weighed and mixed in a plastic bag before adding to the meat batter.

Hygienic aspects of comminuted meat. Comminuted meat is more highly perishable owing to the large surface area exposed to many sources of contamination, the availability of the meat juice, water and nutrients, and the distribution of surface flora during comminution. The flora count varies greatly, presumably reflecting the initial contamination. For example, the count in ground beef is usually less than in minced pork. High counts will arise during comminution if the temperature of the sausage mixture is not properly controlled. It may however be preserved in different ways: adding salt and nitrite, smoking, heat treatment, and adequate storage. The preservative effect of salt is due to its capacity to lower water activity to a level that inhibits bacterial growth. For a longer shelf-life other preservative methods must be used such as the addition of nitrite which has a marked bacteriostatic effect.

If the sausage mixture is kept at a relatively high temperature (above 20°C) before it is filled into casings, vegetative bacteria will grow and cause deterioration. In order to kill most micro-organisms present, emulsion-type sausages must be heat-treated to an internal temperature of 72–75°C. The product shelf-life is thereby significantly extended.

The product must be quickly chilled after heat treatment, in order to prevent growth of any surviving flora. After the final inspection it is ready for sale. Finished products must not come into contact with the floor or any other contaminated materials, especially raw meat or natural casings.

Hygienic Faults Committed during Meat Comminution

The basic objective of meat comminution is to manufacture sausages of standard quality with a desirable hygienic level, and an acceptable shelf-life. For this reason, during comminution all improper handling which can promote growth of micro-organisms must be eliminated. The meat products are not only highly perishable, but may also become the source of foodborne diseases.

Sources of Contamination during Comminution

Unclean machines. Meat batter and sausage mixture represent very good nutritive media for rapid growth of micro-organisms, owing to the presence of the soluble nutrients. If any interruption in production is greater than half an hour the machine must be emptied, the sausage mixture placed in the cooler, and the machine immediately washed. Failure to do so will lead to heavy contamination of the next batch. The contamination will be even more serious if residual material stays in unclean machines overnight or even during the weekend. To eliminate these hazards the machine must be washed immediately after use with hot water. The best method of washing for machines, containers and walls and floors is with a special cleaning machine. When attached to the water supply the machine can increase the water pressure tenfold to make cleaning very effective. Liquid disinfectant may also be added. Hot water (minimum 83°C) gives the best results.

During dismantling of any machine, either for daily washing or routine control of cleanliness, dismantled pieces must not be placed on the floor. They should be put in a plastic (not metallic) tray or vat and after washing hung on a rack, for draining overnight. The next day their cleanliness and sharpness must be checked before use. Water remaining after cleaning must be eliminated before leaving the workshop.

Unhygienic handling. There are many manual operations in a nonautomated production line for comminuted sausages. Unhygienic handling can provoke rapid bacterial growth, which will persist even after hightemperature heat treatment, resulting in a shorter shelf-life. Meat trays, other tools, plastic aprons etc. must never be washed on the floor. Any material falling on the floor must not be reused in further processing. All dirty and contaminated material must be placed in a special watertight non-corrosive bucket, with a lid fastener, to prevent access to unauthorized persons and clearly marked in large red letters "condemned".

HAMBURGER-TYPE PRODUCTS

Hamburger-type products are made form minced meat and fatty tissue to which salt and seasonings are added. Soy is a common additive but curing agents need not be used.

Raw Material

Well-chilled Meat I and Meat II and fatty trimmings are the basic raw materials. All beef, all mutton, all pork or combinations of different meat and fatty tissues in various proportions may be used.

Additives

Extruded soy proteins (ESP) are often used in manufacturing hamburgertype products. They should be hydrated in water (70°C for 35–40 minutes) in the ratio of ESP to water of 1:1, 1:2 or 1:3. Salt, seasonings and dry or fresh onions are the other basic ingredients. If these products are intended for long-term storage some antioxidants may be added.

Manufacturing

Selected meat is minced through a 5-mm, fatty trimmings through a 3-mm, and hydrated ESP through a 3-mm plate. After mincing, the components are mixed with the additives and seasonings until all components are evenly blended. The mixed batter is then ready for forming and shaping. Weight and shape can vary but patties are usually round, weighing 80–120 g and 5–10 mm thick. After freezing, the hamburgers are packed into suitable plastic bags or cardboard boxes, and immediately transported to a storage room.

Beef II or Pork II are used with lower-quality products and ESP is hydrated with more water (1:2, 1:3). Hamburger-type products should contain 10 to 25 percent fat but if hydrated ESP is used up to 30 percent may be added.

Storage and Shelf-life

Hamburger-type products are stored at -20°C to give a shelf-life of 90 days. They may be cooked without thawing by grilling, frying in deep fat or toasting on a hot plate.

COOKED SAUSAGES

Cooked sausages are fine-comminuted sausages representing an important group of meat products. The basic technological aim is to increase the natural WHC of the meat batter to bind considerable quantities of added water. Cooked sausages may be made of uncooked or cooked raw materials.

Cooked Sausages made of Uncooked Raw Materials

These are emulsion-type sausages. The finished products are heat-treated to between 75° and 80°C to achieve desirable organoleptic properties and bacterial stability. They are juicy, tasty, easily chewable and digestible. To achieve such juiciness, about 25 percent water is added to the meat during manufacturing. Phosphate must be added to improve the WHC of the meat proteins. Specific amounts of fatty tissues must also be added to achieve the characteristic consistency and flavour.

Raw material. Many different kinds and qualities of meat, organs and fatty tissues can be used. Beef, veal and pork are more frequently used, but camel, mutton, goat and venison can also be incorporated. Practical experience has shown that more desirable cooked sausages of the emulsion type are manufactured when hot-boned meat is used. In this case phosphate is not necessary, because hot-boned meat has a high WHC in the first three to four hours after slaughter and can emulsify 22 percent more fat than chilled or frozen. If hot-boned meat is not available only well-chilled and/ or frozen meat should be used. The chemical composition, water-binding and fat-emulsifying properties of the raw materials must be known in order to make products of uniform quality. Suitable raw materials are listed in Table 17.

Offals such as heart, tongue, spleen, lungs, stomach, tripe or non-lactating udder are not commonly used in the manufacture of emulsion-type sausages. However, special products of this type may contain smaller quantities of certain offals.

Fatty tissues are also important raw materials. They contribute to the palatability, tenderness and juiciness of sausages. Variation in the ability to emulsify fat is due to the amount of soluble proteins potentially available, and their emulsifying capacity. Ice water is used as a carrier for the curing agents, and improves the extraction of meat protein and the hydration of meat-extender proteins. The amount of added ice water depends on the raw material used. Water can be partly substituted by blood plasma which is an additional source of protein and improves the water-binding and emulsifying properties of the product. Regulations for emulsion-type sausages in many countries limit fat content to a maximum of 30 percent and water content to 40–60 percent in finished products.

Additives. A variety of additives can be used in addition to meat extenders. Phosphates and seasonings (spices, sweetners, glutamates) may be used to create a range of products. If frozen and chilled meat is used, 0.3–0.5 percent phosphate must be added. Salt and pepper form the basis for a cookedsausage seasoning mixture. All other seasonings are only supplementary to them. Ascorbic acid, sodium ascorbate or sodium erythorbate (about 0.4 g per kg) are used in a 10-percent water solution at the end of the comminution.

Casings. Natural or synthetic casings can be used (Table 18) but in both instances they must be firmly filled.

Manufacture. The basic aim in cooked-sausage manufacturing is to make a heat-stable meat batter, a meat "emulsion", consisting mainly of water and protein. Heat treatment transforms it from a viscous form to a rigid and elastic solid structure which can be considered as a protein gel with entrapped fat particles. If the protein gel is strong enough, the fat will not be separated during heat treatment.

Principles of mincing. Mincing completely destroys the meat structure. Meat proteins in the presence of salt, phosphates and cold water are dissolved forming a system consisting of a solution of salt-soluble proteins and muscle and connective tissue particles. When fatty tissue is added the fat particles will be emulsified with the salt-soluble proteins during comminution. The proteins will therefore be able to cover the total fat surface, surrounding each fat particle and so stabilize the emulsion.

During comminution the structured breakdown of meat and fatty tissues occurs and new systems are formed: minced meat, suspension, after addition of salt, phosphate and ice water and emulsion, after addition of fatty tissues and stable sausage mixture after addition of all other additives.

Principles of filling. Cellulose and collagen casings are ready for transferring directly to the filling horn. Salted, natural casings should be flushed thoroughly prior to use, and filled firmly.

Principles of heat treatment. Proteins show a great tendency to bind to and cross-link with each other upon heating. The binding properties of comminuted sausages originate from the heat-induced gelation of meat proteins. Gelation begins at about 43°C and produces a three-dimensional network structure which stabilizes the fat and water in comminuted meat products and binds meat and fat pieces together.

Frequent faults committed during production

- Separation of fat in the form of fat caps or fat pockets develops during heat treatment, due to emulsion breakdown, as a result of improper formulation (too much fatty tissue, especially soft or insufficient emulsifier), incorrect order of addition of raw materials and additives, and overcooking.
- Holes in sausage mixture caused by improper filling.
- Shrinkage of cooked sausages caused by improper smoking and/or heat treatment (temperature too high and RH too low).
- Difficult peeling of casings, can be the result of too much connective tissue in the formulation.
- Uneven colour of smoked sausage surface (unsmoked spots) occurs at places where links touch each other.
- High porosity of sausage mixture due to entrapped air.

Different products. The best-known emulsion-type cooked sausages made of uncooked raw materials are frankfurter, Vienna sausage, hot dog, knackwurst, safalada, bologna and veal sausage.

Frankfurter-type sausages are sold under many names: frankfurters, wieners (braided in groups of links), Vienna-style sausage (twisted into a chain of links), hot dog, knacker, etc. All these types normally contain more beef than pork (1.5:1). The choice of raw material largely depends on the availability of the meat, the eating habits of the consumers and religious practices. In general, the lean-meat content varies between 25 and 40 percent, the average fat content is 25-30 percent, and 10-30 percent water is added. In low-quality frankfurters, 10 percent of the lean meat can be replaced by pork skin or organs. To maximize yield part of the fatty tissues and part of the water may be pre-emulsified (10–15 percent).

If a high-speed cutter is not available, the first manufacturing step is grinding. The meat and trimmings must be ground through a 3-mm, and fatty tissue through a 5-mm plate.

The temperature of all raw material after grinding should not exceed 3°C. If a high-speed cutter is available, then a grinder is not necessary as the sausage mixture can be made in the cutter. Attention should be paid to the order in which raw materials are added (meat, curing agents, ice water, meat extenders, fatty tissues and seasonings). Water must be added early enough to ensure protein extraction. It is important to control the temperature during comminution, not only for the extraction of salt-soluble proteins but also for the hydration of meat-extender proteins. The final temperature of the sausage mixture should not exceed 18°C if pork fatty tissues are used, and 2°C if beef fatty tissues are used.

After filling in natural, synthetic, edible or inedible casings and linking, products should be smoked immediately. They are hung on sticks and placed on smokehouse racks. Links should be separated on the stick so as not to touch each other, otherwise smoking and colour development will not be uniform. Surfaces in contact with the stick or with other links do not dry out sufficiently, favouring the growth of micro-organisms forming slimes, especially in warm conditions. This can be eliminated by shifting the links slightly at some time during smoking to expose the surfaces which have been in contact with the stick. This can be minimized by using "T" profile metal sticks with apertures.

Smoking imparts an appetizing flavour and develops a desirable goldbrown smoked colour. It also contributes to the development of a smooth surface or skin beneath the cellulose casing, that facilitates peeling of the casing prior to packaging. Smoking lasts from one to three hours, depending on desirable colour, temperature and humidity, and type of smokehouse employed. A high smoke concentration and a rather high temperature are desirable during the final period of smoking. Smoking and heat processing start at 54–57°C and the temperature is gradually increased (5.5°C every 15 minutes), until it reaches 82°C. Smoking continues until an internal temperature of 70–75°C is reached. The next step is cold showering until an internal temperature of 32°C maximum is reached. The product is then placed in a cooler.

Vienna sausages are manufactured using beef and pork trimmings with a maximum quantity of water (28–30 percent) and meat extenders, tied in similar lengths to frankfurters (12–14 cm).

Hot dogs have the same formulation as Vienna sausages but are tied longer than any other emulsion-type small-diameter cooked sausages (16–18 cm).

Safaladas/knackwursts have a frankfurter composition but the structure is finer than any other cooked sausage. It is obtained by passing the sausage mixture twice through an emulsion mill, after being manufactured in the cutter.

Bologna sausage is a typical emulsion-type cooked sausage filled into large casings. Curing is as in frankfurters but less water is added. Long bologna is filled into beef middles or weasand, large bologna into beef bungs, and ring bologna into beef rounds. Cellulose casings of corresponding size may also be used. Large bologna must be supported either by wrapping it two or more times with twine and hanging it on a loop in the twine or by supporting it in a stockinette bag. Bologna requires longer smoking (three hours for long and five for large) and cooking than frankfurters.

All parts of the bologna must reach 71°C during cooking or the inside of the sausage may become discoloured before it is sold to the consumer. After cooking the bologna is chilled in cold water, superface dried, then placed in a cooler.

Shelf-life of cooked sausages is about four to five days, or up to ten days in retail packages (vacuum) held at 10–15°C or as high as 12 to 14 months in cans or jars.

Small and medium-diameter cooked sausages are reheated before eating in order to regain their firmness and juiciness. Reheating is done in boiling water for about ten minutes (or by toasting or grilling). Cooked sausages are eaten while hot with baked goods (rolls, buns), mustard, horseradish, ketchup etc. Large-diameter sausages are cut into slices as a cold starter or used in sandwiches, Russian salads and similar light meals.

Cooked Sausages made of Precooked Raw Material

These are sausages made from edible offal and lower-quality meat and have a specific flavour. The raw materials are precooked, minced, mixed, filled into casings and finally heat-treated. Fine-comminuted sausages are processed in the cutter and coarse-comminuted in the grinder. There are also sausages with a mixture of both fine- and coarse-comminuted pieces of meat, tongue, liver, fatty tissue (all these components are most often cured). Uncooked cured blood (2 percent nitrite salt) is often added and precooked pigskin is also very often used, except with all-beef products in which tendons and other connective tissues are used instead.

Raw materials are all edible offals, head meat, lower-quality meat, fatty tissues and pigskin. All components except pigskin are fully cooked. Pigskin is partially cooked until soft enough to be ground in the grinder. Its function is to give the finished product its typical firmness.

Precooking is done in boiling water with sufficient water in the cooking vats to cover all the material. Cooking time depends on the size of pieces. Fully cooked raw materials are comminuted without delay. If comminution is done in the cutter then no blending is necessary.

Additives used are nitrite salt, caseinate, blood-plasma powder, egg white, gelatine and seasonings. All except the seasonings can be added at a rate of up to 2 percent.

Casings. Natural and synthetic casings can be used but in countries with a hot climate, if continuous refrigerated storage during retail sale is not possible, priority should be given to synthetic casings. Such casings are also more convenient because they can be exposed to higher temperatures (up to 115°C retortable sausage) during heat treatment giving a longer shelf-life.

Composition. There are four types of precooked sausages which differ in composition and manufacturing method:

- liver paste;
- liver sausage;
- blood sausage; and
- cooked sausage with high collagen content.

In all formulations meat can be substituted with other edible offal. Liver paste is fine-comminuted with high spreadability.

Liver sausage contains 10–20 percent liver, usually coarse-ground, and other edible offals.

Blood sausage contains 10–20 percent whole blood with nitrite salt (not precooked). Other components are precooked meat, edible offals, fatty tissue and pigskin. Fatty tissue must be cooked sufficiently in order to separate fat with a low melting point.

Components can be coarse- or fine-comminuted as in other precooked sausages. This type of sausage has a firm consistency due to swollen connective tissue components and gelatinized collagen.

Cooked sausage with high-collagen content resembles coarse-textured liver sausage but has more connective tissue (collagen and elastic of pigskin and beef tendons). Sometimes up to 10 percent dissolved gelatine solution is used to produce a sausage described as aspic in casing.

Heat treatment. The mixture is filled into the casings while still hot and the sausages are immediately heat-treated. Cooking time should be set according to the type (natural or synthetic) and diameter of casing. Natural casings exposed to temperatures higher than 85°C will burst, therefore their shelf-life is shorter (up to ten days at 5°C). Synthetic casings can be cooked in boiling water for about 90 minutes, or even in an autoclave at 110°C for 60 minutes. If they are kept at temperatures below 10°C they will have a shelf-life of one month.

Frequent faults committed during production

- Undercooking resulting in a shorter shelf-life
- Overcooking may result in an oversoft consistency

- Insufficient connective tissue will also result in a soft consistency, and the filling will fall apart
- Fat separation may occur in the liver paste if insufficient emulsifier is added.

The most serious fault is not to fill the casings immediately and not to heat-treat them immediately. In this case the product can spoil even before the heat treatment.

Different products. There are many products of this type which differ in the raw materials, grade of comminution and seasonings. Sometimes cooked rice is added to the sausage mixture.

Eating method. Liver paste is spread on bread or is used to make sandwiches. Other types of sausage are cut into slices and consumed cold.

DRY SAUSAGES

Dry sausages are specific coarse-comminuted meat products whose successful manufacture depends upon bacterial fermentation. At some stages of processing, usually during smoking, these sausages are deliberately held at temperatures which encourage bacterial growth and fermentation. Dried sausages should never exceed 30°C at any stage as this would stop growth. The useful organisms responsible for desirable fermentation (lactic acid-producing bacteria) originate from the natural flora of the meat, processing equipment and the plant environment. Fermentation causes a characteristic tangy flavour to develop, resulting from the accumulation of lactic acid and many other fermentation compounds. The pH usually falls to 4.8–5.4. Another basic processing preservative step is dehydration achieved by keeping the product under controlled temperature and air humidity (drying and ripening).

One of the distinctive features of dry sausages is that they are processed uncooked. The low pH (high acidity), low-water content and high-salt content extend their shelf-life. Some but not all dry sausages are smoked. The best known are dry pork sausages, dry beef sausages, mixed dry sausages, summer sausages and salamis.

Raw Material

In general, dry sausages are composed of two-thirds meat and one-third fatty tissues to which curing agents and spices are added. Meat I and Meat II of all species of slaughter animals can be used, including camel, donkey and horse meat, but rarely mutton, goat or venison. Trimmings are not used owing to their softness, neither are shanks or head or boar meat. Pork jowl and back fat, beef external fatty tissues and humps are used as fatty components.

Only well-chilled (-1° to 0°C) and/or frozen (-1° to -18°C) meat is used. Frozen meat is thawed for 48 hours on slanting racks to allow the drip to run off. The meat is chopped with sharp regular cuts, without crunching. The temperature of meat and fatty tissues should be adjusted to stay in the range of -2° to 5°C during chopping so that temperatures in the filler will be between-1° and -3°C. If the temperature in the filler is higher, drops of fat are deposited on the interior walls of the filler horn. During further filling they are pressed into the casing and will grease the interior of the casings lowering their porosity, making smoking, drying and ripening more difficult.

Main Additives

Salt (28–37 g per kg) is used to prevent the growth of many undesirable aerobes, favouring the growth of non-spoilage halophile and halotolerant bacteria. Salt also extracts the salt-soluble proteins to form a protein gel which binds the pieces of meat and improves sliceability of the finished product.

Curing substances are used in the form of nitrite salt or a dry mix of common salt with 0.6 percent nitrite salt and nitrate (0.3–0.5 g nitrate per kg).

Sugars speed acidification and are transformed into lactic acid by certain bacteria. Dextrose is the most commonly used sugar (8–10 g per kg) but can be replaced by saccharose.

Spices. White and black pepper (0.5–3 g per kg), ground or crushed, are the most frequently used spices. Fresh crushed garlic, paprika, cardamom, mustard etc. are also used. In some countries wine is added to improve the flavour. Some antioxidative substances may also be added.

Starter cultures. To overcome the problems associated with bacterial fermentation, starter cultures of selected lactic acid-producing flora have recently been used. The starter culture provides a predominant flora of the desired bacteria (*Micrococcus, Pediococcus cerevisiae* etc.) in the sausage mixture, and fermentation is initiated within a minimum time.

Casings

These may be natural or synthetic. Natural casings are usually small intestine of hog (for small-diameter dry sausages), beef small intestine (for middle-diameter dry sausages), and large hog or beef intestine (for largediameter dry sausages). As for salami (diameter more than 40 mm) and summer sausages the most convenient casings are small horse intestine. The best synthetic casings are the so-called "dry sausage fibrous casings", as they adhere very well to the product as it shrinks during drying.

Composition of Some Popular Dry Sausages

Like all other sausages there is much variation in the composition of dry sausages and salamis. A common factor is that they do not contain any added cereal, ice or water.

Manufacture

In the successful manufacture of many varieties of dry sausages and salami, a great deal of technical skill is indispensable. In manufacturing coarsely chopped sausages the grinder and mixer are used but for less coarsely chopped sausages only the cutter is used.

In the first case, meat and fatty tissues pass through the grinder (beef 15-mm, pork 2–6-mm and fatty tissues 8–10-mm plate) and the sausage mixture is made in the mixer. In the second case, chopping and mixing are done in the cutter at slow speed. After adding the curing agents the meat is chopped to the required size then fatty tissues and all other components are added. The mixture is firmer and without residual air if a vacuum is applied. The sausage mixture must be firmly filled into casings. The casings are punctured with small needles over the entire surface to allow entrapped air to escape. Tied sausages are hung on sticks and transferred into a curing room. Curing time depends on the sausage diameter, three days for diameters up to 3 cm; and five days for larger. During this period, the cured meat colour is developed and fermentation is initiated.

After curing, sausages are transferred to a drying room. The rate of drying is controlled by keeping the products between narrow limits of both temperature and RH. Too rapid drying will result in the formation of an outer crust on the sausages, which will retard or stop internal drying. If the drying rate is too slow and RH is too high, then surface mould, yeast and bacterial growth are excessive. During drying and ripening, flavour develops, texture changes and the product hardens. Degree of drying can vary according to local preferences. Ripening can either follow or precede cold-smoking, depending on the particular product.

Dry sausages must be stored in an environment in which temperature and humidity do not provoke overdrying. They are ready for sale immediately after production. All dry sausages may be sliced and packed under vacuum in different consumer sizes. They are consumed sliced, as starters or in sandwiches.

Only complete fulfilment of all technological parameters at all stages of production can guarantee the desired quality of finished products. Dry sausages, like dry hams, are high-quality meat products. The main characteristics of dry sausages are agreeable bouquet, i.e. flavour and taste of a well-matured cured-meat product, attractive colour, good sliceability and long shelf-life. In order to achieve all these, sufficient time must be allowed of at least a month for small-diameter sausages and three to six months for large-diameter sausages and salami. Organoleptic tests and weight-loss control may confirm the end of ripening. Water content of finished dry sausages (of any diameter) should be between 25 and 30 percent. Sausages should not be overdried or they are hard to chew and less acceptable to consumers.

Semi-dry Sausages

They are produced by quick drying without ripening. Depending upon the diameter, this lasts from two to three weeks. In order to shorten the process, reducing agents (such as Glucono-delta-Lactone) and a starter culture must be added to the meat mixture. Weight loss is lower than in dry sausages, so the water content of the finished product is always greater than 40 percent, resulting in a shorter shelf-life, a sour taste and a poor flavour. Sliceability is also owing to reduced binding of meat and fatty tissues. Semi-dry sausages have a maximum shelf-life of one month at room temperature.

Frequent Faults Committed during Production of Dry Sausages

Creases or detachment appear in sausages with synthetic casings if the filling dries at a faster rate than the casing. Small creases become larger as drying continues, and overdrying causes detachment of the casing. This can be prevented by using more elastic casings and by a slower drying rate.

Crust formation results from rapid drying, especially at the initial stage, and can be detected by careful palpation. If the crust is not firm a preventive measure is to increase the humidity in the drying room to soften and rehydrate the sausage.

Greasy casing results from using soft fatty tissues or from overheating the sausage mixture. Melted fat, under pressure during filling, greases the interior walls of the casing reducing its porosity and preventing normal water migration from inside the sausage. The consequence is a soft sour sausage. If the melted fat penetrates the casing walls, then the product also becomes greasy.

Sausage mould, usually greenish in appearance, is the sign of contaminated equipment and workshop. It is eliminated by using good antifungicidal agents in cleaning operations.

Sausage sliminess is the result of heavy micro-organism growth on the surface of the sausage casing, encouraged by too high a temperature and air humidity. Partial sliminess can be removed by washing in salty water, followed by dripping and more intensive and dense smoking. This fault may appear when the product is stored in cardboard boxes for a long time.

Sour sausage, being invisible, is unfortunately usually detected only at the end of drying during the final control. It has a sour taste, semi-rigid consistency and the periphery is darker than the centre. It is the result of the intensive growth of lacto-acid bacteria, feeding on the added sugar. Preventive measures are to decrease the added sugar and increase the added salt. Such products must be properly dried and ripened.

Sausage porosity usually appears in large-diameter sausages. Air in pores causes decomposition of fat, leading to changes in colour and taste.

Poor sliceability is the result of insufficient curing and salting, uneven distribution of curing agents, and/or short curing. The lack of salt-soluble proteins reduces binding between meat and fat particles.

Blown-up sausages due to gas produced by sudden micro-organism growth result from contamination of raw materials, equipment, tools and unhygienic production.

Dry sausages are a very nutritive meat product which may be eaten immediately after production, without any additional heat treatment. They are sold entire or packed in pieces in plastic bags. All dry sausages can be sliced and packed under vacuum, in various consumer sizes. Dry sausages are usually covered in very thin slices in sandwiches, or eaten as starters.

MEAT EXTENDERS

Meat extenders are usually protein additives, defined as non-meat proteins.

A wide variety of meat extenders are available for use in emulsion-type sausages to improve consistency and emulsifying and water-binding capacities. They can also serve to enhance protein content, improve processing yields and reduce formulation costs. The most important meat extenders are soy proteins, milk proteins, starch, flours and yeast. The permitted maximum amount of meat extenders in sausage production is usually 3.5 percent and is strictly regulated by law in certain countries.

Soy Proteins

These are table proteins from soybeans in the form of flours, grits, concentrates and isolates, texturized, untexturized and extruded.

Soy flour is a screened, graded product, obtained after extracting most of the oil from dehulled soybeans.

Soy grits are particles of larger size, described as coarse, medium and fine.

Soy flours and grits may be defatted, low-fat, or high-fat. Owing to their functional properties (emulsion stabilization, and fat- and juice-binding), soybean flour and grits are used in the production of all emulsion-type sausages, meat patties and canned-meat products. However, both flour and grits give a slightly bitter taste to meat products, which limits their use (up to 2 percent).

Soy protein concentrates are high-protein products (not less than 70 percent). They possess good water-absorption abilities and are used in emulsion-type sausages, luncheon loaves and meat patties (2.5–3 percent).

Soy protein isolates are the most refined form of soy proteins, without water-soluble sugars and other insoluble components. Owing to their highprotein content (minimum 97 percent) they have excellent dispersing, emulsifying, gelling and water-and fat-binding properties. They are used in production of bologna, miscellaneous sausages, canned and dietary meat products.

Textured soy proteins are available in a variety of shapes, sizes, flavours (beef, pork, ham, chicken etc.), textures and colours. They are used as extenders in comminuted sausages, meat patties and canned-meat products. Some physical properties and typical chemical composition of different soy protein products are given in Table 12.

Application of Soy Proteins

Soy proteins and meat proteins interact differently with salt. While salt aids the extraction of salt-soluble proteins from meat, it has the opposite effect on soy proteins since it prevents their hydration. It is important therefore to add meat, soy proteins and salt in the correct order. Salt should be added only after hydration of soy proteins has been completed. For complete hydration sufficient water must be available and enough time allowed. The soy product/ water ratio should be 1:4 or 1:5, with a minimum of three minutes' chopping with the knives and cutter at the highest speed. Soy proteins may be used in sausage production in dry gel and emulsion forms.

Dry form. Meat is placed in the bowl at slow speed and soy product is added. The correct amount of water is added at the highest speed to allow complete hydration of the soy proteins. When the mix is homogenous, phosphate is added and finally nitrite salt.

Gel. Using soy gel ensures that the soy proteins are completely hydrated. To 15 kg of dry soy product, 60 kg of water is added. Complete hydration can be seen by the appearance of the gel consistency. Such gel, if used in sausage production, must be added to meat batter before fatty tissues and other additives. The remaining gel is stored in a cooler and, if not required until the next day, 2 percent salt is added.

Emulsion. Soy products can be added to emulsion-type sausages in the form of a pre-processed soy emulsion. These can be produced by processing with cold or hot water, depending on whether fatty tissues are used. The ratio between soy product, fatty tissues and water is 1:5:5 for firm and 1:4:4 for soft fatty tissues. In both cases, soy product is placed in a bowl, water is added, and the mix is comminuted to complete hydration (minimum three minutes). Fatty tissue is added and chopping is continued at the highest speed. For better bacterial stability 2 percent salt must be added. Coldprocessed emulsion can be used the same day, but hot-processed emulsion must be chilled overnight in flat trays in layers not more than 10 cm thick.

Milk Proteins

These are casein, lactalbumin, lactoglobulin and other serum proteins. Casein is one of the most important proteins, used as an emulsifying agent in sausage manufacturing as caseinate (dry milk-protein isolate). It is a fine white powder with neutral taste and a protein content of approximately 94 percent. Unlike the coagulable soy proteins, egg albumen and blood plasma, milk protein does not coagulate, shrink or form a gel while heating.

If caseinate is added to meat before salt addition, it will surround free fat particles during comminution. Thus, the binding capacity of meat saltsoluble proteins may be enhanced to form a more stable emulsion. In sausage manufacture, caseinate may be used in three ways:

Prefabricated caseinate emulsion. When making this emulsion, the order of addition, the necessary emulsifying time and temperature, and the protein:fat:water ratio are of major importance.

After the fatty tissue has been comminuted to a fine paste, dry caseinate must be added immediately, followed by all the hot water. After four to six minutes of comminution the emulsion is ready. During the last few rotations 1.5 percent salt is added. For high-emulsion stability it is necessary to reach a minimum emulsifying temperature of 45°C for pork and 50°C for beef fatty tissues. The emulsion must be chilled overnight in flat trays. Next day the emulsion may be added to the meat batter, together with fatty tissues.

The usual caseinate:fat:water ratio for the lower-viscosity caseinate (EM-6) is 1:5:5 and for high viscosity (HV) is 1:8:8. If cooked pork skin is used, the ratio of caseinate:fat:water and skin is 1:8:8:2 for sausages.

Dry powder caseinate may be used when prefabricated emulsion or gel are unavailable. As this is absorbed at the fat-water interface it is absolutely necessary to add caseinate before fatty tissues to ensure that the emulsion is heat stable.

Prefabricated gel is a concentrated solution of caseinate in water. Optimum results can only be expected when the caseinate is completely dissolved in water (1:8). It is recommended to moisten caseinate in the cutter with an equal amount of ice (1:1). After a few minutes of comminution the remaining seven parts of water are added. The gel is used in sausage formulations containing relatively small quantities of fat and in coarse comminuted sausages.

Starch

This is a complex sugar of plant origin. The basic technological function of starch is to absorb released water and juice during heat treatment. The most frequently used starches are wheat, maize, potato, rice and manioc. Starch is used in doses up to 4 percent. During heat treatment of sausages, added starch binds part of the free water and swells, thus decreasing weight losses.

Flour

Wheat flour is the most commonly used binder following hydration. Approximate chemical composition is starch 65–75 percent, gluten 8–14 percent, and water 12–16 percent. If flour is added in sausage formulation, care must be taken to prevent quick decomposition of the sausage mixture. Owing to the activity of flour enzymes, flour is commonly used in lowquality sausages (up to 4 percent).

Yeast

Brewer's inactivated yeast with the bitter taste removed is frequently used. It is a fine yellow-grey or grey-pink powder, with distinctive taste and smell. Approximate chemical composition of yeast is: 53 percent proteins, 36 percent sugars and 3 percent water.

In the meat industry the most important application of yeast is based on the ability of its proteins to emulsify fat. A heat-stable emulsion is obtained if the yeast, water and fat ratio is not more than 1:5:5. In comparison with soy isolate and caseinate, yeast proteins have a lower ability to emulsify tallow. A heat-stable emulsion with tallow is only obtained with a ratio of 1:1.3:1.3.

Yeast can be used as an emulsifier in cooked and in canned chopped meat products and can easily replace caseinate in paste production. Yeast used in these products may improve their flavour. The content of free amino-acids improves the flavour of the meat product. Yeast also intensifies the flavour of certain seasonings, especially those with hot components.

Yeast Extract

This is obtained from special yeasts cultivated on cane-sugar nutrient media. It contains many proteins, amino-acids and B-complex vitamins. It is used for improving the flavour of canned meat.

CASINGS

Casings are special cylindrical containers used to protect sausages and various meat products. Since sausages are comminuted products they must be placed in some type of forming device to give them shape, to hold them together during further processing and for protection. Casings may be natural or artificial.

Natural Casings

These are derived almost exclusively from the gastro-intestinal tracts of swine, cattle, sheep, goats and horses. Hog casings are prepared from stomach, small intestine (smalls), large intestine (middles) and terminal end (colon) of the large intestine (bung). Beef casings may be from the oesophagus (weasands), small intestine (rounds), large intestine (middles), bung and bladder. The intestines of sheep and goats are used to produce casings primarily for fresh sausages, frankfurters, hot dogs etc. Natural casings can be classified as narrow, medium and wide.

Treatment and storage. Intestines intended for use as casings for sausages must be immediately processed after evisceration. First they are emptied and well flushed. Pig, sheep and goat small intestines are scraped thoroughly without inversion to remove the exterior (serous) and interior parts (mucous membrane), preserving the middle elastic muscle. The same method is applied to the large intestines of all animals.

Well-scraped, flushed and drained intestines are well sprinkled with common salt (dry salting) or saturated salt solution (wet salting), spun and tied into intestine completes. They are arranged into plastic casks. The top of the intestines is pressed down and salted again in order to eliminate air contact. These casks must be firmly closed and stored in a dark cooler. If intestines are prepared for long storage (more than three months), they must be salted with a mixture of salt and 0.2 percent antioxidant.

Large intestines and bladders can be preserved by means of air-drying. In this case, after scraping, flushing and a short draining intestines are blown up in order to eliminate the remaining water and tied at both ends in hanks, for quicker air-drying. They should be hung in a sunny, draughty place. When well dried they are sorted according to diameter and packed. Such dry casings do not require storage in a cooler.

Application. All salted natural casings should be thoroughly flushed inside with running water the morning before use. They are then dipped into warm water to regenerate their elasticity (small intestines 10–20 minutes and large intestines 30–60 minutes) and drained for a short time before use.

Artificial Casings

The advantages of these are uniform cylindrical shape, a range of specific diameters, suitable tensile strength, resistance to damage, ease of use, variety of sizes and low microbial level. They can be filled uniformly and linked either by hand or machine into regular lengths.

Four Types of Artificial Casings are available:

Cellulose casings are prepared from cotton linters in sizes ranging from 1.5 to 15 cm. They are easy to handle and fill, possess a high degree of resistance and are permeable to smoke when moist. The degree of permeability decreases as the casing dries during processing. Small cellulose casings are used for skinless frankfurters or wieners, skinless smoked link sausages and many other small sausage products (cocktail sausages). Small cellulose casings are also available in shirred form.

Large cellulose casings are available in three types: regular, highstretched and large. They are used in making all types of sausages and smoked meat. Cellulose casings are inedible and must be removed before consumption.

Inedible and edible collagen casings are regenerated from collagen extracted from skins and hides. The inedible collagen casings combine some of the advantages of both natural and artificial casings, especially their strength, uniformity and shrink characteristics. They must be removed prior to consumption. Edible collagen casings are mostly used for fresh pork sausages and frankfurters.

Synthetic casings and bags are impermeable to smoke and moisture. They are used with unsmoked products such as liver sausages or products which are heat-processed in water or a retort.

Tying of Casings

Artificial casings are delivered either tied at one end with loops prepared in the factory or in hanks. As casings in hanks can be easily stored and their length and the method of tying can be chosen at will, they are well suited to small-scale production. One end of the casing is submerged in cold water (one to two minutes), then tied either in a "doll" or "bow" knot. To make a "doll" knot, place a string on the casing and twist around, slip the thumb and forefinger on the string 1 cm toward the end of the casing, pass the string through the bight and set the knot, tighten the string, and prepare the loop and tighten with a double knot. When tying a "bow" knot, casings need not be submerged except the end of every dry casing. The method is to place string on casing and tighten by means of a simple knot, fan-shape casing end with knot in the middle, and tie knot underneath the wing (the tied casing has no loop). This method of tying prevents slipping off.

5

Bacterial Growth on Meat and Meat Products

The total viable count of bacteria (TVC) expressed as organisms/cm^2 or as organisms/g on fresh meat or a meat product sets a limit to its shelf-life. Meat will "spoil" with TVC at $10^6/cm^2$because of off-odours. Slime and discoloration appear at $10^8/cm^2$. The main factors determining the time taken for the TVC to reach these levels are the initial count due to contamination during slaughtering and processing, further contamination during storage, temperature, pH and relative humidity. An example of how the level of contamination affects shelf-life.

After cleanliness, keeping meat products cold is the second most important requirement in order to achieve a desirable shelf-life. Microorganisms rapidly proliferate at elevated temperatures and slime development is a definite visual sign of microbial growth. The importance of temperature in the control of microbial growth.

Bacteria relevant to meat, meat products and other food are divided into three groups according to the temperature range within which they can grow: mesophiles 10–45°C, psychrophiles 0–28°C and psychrotrophs 10– 45°C, or slow growth at 0–10°C. Mesophiles will not grow below 10°C but psychrotrophs, of which Pseudomonas are the more important, will grow down to 0°C. The nearer to 0°C the storage temperature the slower the growth of the spoilage bacteria and the longer the shelf-life.

Some bacteria cause product spoilage, others cause food poisoning. The former limit product shelf-life but the latter cause illness. Almost all foodpoisoning bacteria are mesophiles so refrigeration below 10°C offers good protection. Many mesophiles cause spoilage, but since meat is refrigerated most spoilage is due to psychrophiles. Storing meat at temperatures close to 0°C will inhibit the growth of pyschrotrophs. Shelf-life will be extended by avoiding contamination through good hygiene practices.

Aerobic bacteria have an absolute requirement for oxygen which limits their growth to the meat surface. Anaerobic bacteria grow within the meat as they need the absence of oxygen. Facultative anaerobes can grow slowly within oxygen but grow better in its presence. Food-poisoning bacteria are anaerobes and facultative anaerobes. The most important spoilage bacteria (*Pseudomonas* spp.) are aerobic.

Water is required by micro-organisms so reducing the water available below the optimum level will prolong shelf-life. If meat is stored at a relative humidity (RH) below 95 percent, moisture will be lost from the surface. Since most spoilage bacteria, being aerobic, can grow only on the surface, this surface drying will extend the shelf-life. Moulds (fungi) are able to grow in drier conditions than bacteria so that desiccation has a selective effect on microbial growth.

Meat pH is the level of acidity in meat. Stored sugars are broken down to lactic acid. In living muscle it is near 7.0 (above this is alkaline, below is acid). It falls to 5.4–5.6 within 24 hours. High final-pH values result when animals are exhausted at slaughter, for instance because of fighting in lairage or transport. Spoilage bacteria multiply rapidly on high-pH meat and shorten the shelf-life. Exhausted animals should be rested before slaughter.

A high TVC resulting from severe contamination during slaughter or processing will shorten the shelf-life even in ideal conditions. It also indicates poor hygiene so that contamination with food-poisoning bacteria is likely.

HYGIENE RULES FOR MARKETING CHILLED MEAT CUTS

Chilled meat is usually kept for the sale in refrigerated display cabinets, either unwrapped or portioned and packaged for self-service outlets. Refrigerated display cabinets may have fan-assisted convection and/or natural convection. Fan-assisted types are better able to maintain a lower temperature as they are less affected by draughts. Cabinets should be stacked to maintain a good air flow around all meat.

Do not store or display unwrapped cooked and raw meat together. Use separate refrigerators, display cabinets etc. to avoid cross-contamination. Raw-meat exudate on to cooked meat gives an explosive bacterial growth.

Simple packaging of fresh meat with plastic foil has become very popular with the availability of suitable and inexpensive film. The main objective of simple packaging is to provide hygienically protected portioned meat for self-service retail outlets.

But the meat portions must also satisfy the customers' preference for bright red fresh meat. This colour is due to the pigment myoglobin loosely binding oxygen to form oxymyoglobin. For this colour to develop and be maintained, the wrapping film must have a high-oxygen permeability. To avoid desiccation of the cut surface, the film should have a low-moisture permeability. After a time the cut surface becomes more brown as a result of myoglobin binding the oxygen more tightly to form metmyoglobin. This may take up to three days depending on the temperature, the number of bacteria and other conditions.

Simple packaging for retail sale in self-service outlets usually involves placing the meat portion in a plastic tray and overwrapping with a clear plastic film. Plastic trays are more hygienic than cardboard. The portions cut should be based on local demand and only a day's sales should be cut at a time.

The principal object of this type of simple packaging from a hygiene point of view is to reduce contamination from airborne micro-organisms. High standards of hygiene are required in the cutting and packaging operations. On large pieces of meat the bacteria mainly colonize the outer surfaces. When meat is cut even with a clean knife they will be spread on to the freshly cut moist surface and multiply rapidly. This is not an argument for relaxing hygiene standards, rather it underlines the need not to add to the bacterial load by further contamination.

All surfaces and tools in the cutting and packaging room must be kept thoroughly clean. Packaging materials should be stored in hygienic conditions protected from dust and attack from insects or vermin. It is most important that personnel involved in cutting and packaging pay particular attention to personal hygiene as they are the most likely source of food-poisoning pathogens which may survive better in the package environment than on unpackaged meat. This is in part due to the packaging preventing surface desiccation. The moist surface favours bacterial growth as does the high relative humidity that builds up within the pack.

It is important to retard bacterial growth by maintaining a low temperature during the display life of the packs. Overwrapping actually increases the meat temperature as the layer of trapped air acts as an insulator. Heat generated by light warms the upper surface. Meat should be thoroughly cooled before packaging to help maintain a low temperature during its display life.

Mincing meat spreads bacteria on the surface all through the meat which therefore has a shorter shelf-life than cuts. Mince may be packaged and overwrapped but the mincer must be kept scrupulously clean and the packs kept well chilled. Only small quantities of mince should be prepared at a time.

Cooked meats, which typically have much lower bacteria counts than fresh, are more open to attack from airborne micro-organisms as these will be faced with little competition. Packaging is therefore particularly beneficial in preventing this type of contamination for cooked meats.

Bacteria introduced during cutting and packaging face little competition and may be of the food-poisoning type if personal hygiene is poor.

If very high standards of hygiene cannot be maintained then a pasteurizing treatment after packaging will be necessary. Even this, however, will not guarantee destroying *Bacillus* and *Clostridium* spp. if these have been introduced.

COOKING METHODS FOR DIFFERENT MEAT CUTS

Primarily because of natural tenderness or lack of tenderness, different cooking procedures are utilized to prepare the various cuts of meat correctly. Tender cuts are best cooked with dry heat, as by broiling, roasting or pan broiling. Less tender cuts are tenderized by cooking with moist heat. Connective tissue is softened and made tender by cooking slowly in moisture.

Temperature control is important in meat cookery. Meat loses moisture, fat and other substances such as soluble proteins during cooking.

Cooking losses can be minimized by controlling the cooking temperature and the final internal temperature of the meat. Higher oven and higher internal temperatures increases shrinkage.

Whenever possible a meat thermometer should be used to determine accurately the degree of doneness of meat. Time and temperature guides can be used to ascertain doneness, but cooking time is affected by fat, bone and moisture content and the shape and size of the cut. The basic types of meat cookery follow.

Broiling

Broiling is recommended for all tender cuts and for best results:

- Set the oven for broiling
- Place thin cuts of meat on a rack at a distance from the heat equal to two times the thickness of the cut plus 2.5 cm
- Broil steaks, chops or patties for approximately one-half the desired cooking time before turning
- Season and serve at once.

Pan-broiling

Pan-broiling is recommended for tender cuts suitable for broiling. For best results:

- Place meat in a hot frying-pan or on a griddle
- Do not add fat or water
- Cook slowly over moderate heat, turning occasionally
- Pour off or remove fat as it accumulates

- Brown meat on both sides
- Avoid overcooking.

Roasting

Roasting is recommended for large, tender cuts. Some beef cuts suitable for roasting are rib and top sirloin roasts. For best results:

- Season with salt and pepper as desired
- Place the meat, fat side up, on a rack in an open shallow roasting-pan
- Insert a meat thermometer so that the bulb is in the centre of the largest muscle without touching bone.
- Add no water and do not cover
- Roast at oven temperature of 176°C to desired internal temperature.

Meats are usually cooked to degrees of doneness as follows:

- Rare 60°C
- Medium 71°C
- Well done 77°C

Pan-frying

Pan-frying is usually recommended for tender cuts 2.5 cm thick or less. For best results:

- Place meat in a hot frying-pan or on a griddle
- Fat may be added
- Cook slowly over moderate heat, turning occasionally
- Allow fat to accumulate
- Brown meat on both sides
- Avoid overcooking.

Braising

This method is best used for less tender cuts such as beef round or chuck steak, pot roast, stew or short ribs. For best results:

- Use a heavy pan
- If desired, brown meat slowly on all sides with sufficient fat to keep meat from sticking
- Season with salt, pepper, herbs or spices
- Add a small amount of liquid
- Cover tightly
- Cook slowly over low heat on a stove burner or in a moderate oven until meat is tender.

Braising with large cuts is often called pot-roasting and with thin cuts may be known as Swissing.

PORK CUTTING

Halving is done immediately after the animal has been dressed and every effort should be made to saw the carcass into equal sides through the centre of the backbone. The side to be cut should be laid on the cutting table with the inside up.

The primal cuts of pork are: ham, fore-end or forequarter, loin and belly.

Hind foot. The hind foot is removed by sawing through the hock joint at a right angle to the long axis of the leg.

Ham. The ham may be removed in several ways to make either long-cut or short-cut hams. One procedure (short-cut) is to locate the division between the second and third (or the third and fourth) sacral vertebrae and saw perpendicularly to the long axis of the ham. After the bone has been severed with the saw, the knife is used to complete the removal of the ham. The ham is further trimmed by removal of the tail bone on one side and the flank on the other side. Commonly a skinned ham is produced by removal of three-fourths of the skin and fat from the rump end. For the production of special cured dried hams the skin is left on. The cutting procedure of the ham is as follows. Remove tail bone and aitch bone and cut the rump off. Peel back the rind and associated fat to expose the topside muscle on the interior side of the leg. Separate the topside by following the natural seam between it and the silverside (outside portion of leg) and thick flank (front position of leg). The topside can then be sliced into steaks. This produces between five and six lean steaks depending on the thickness and weight required by the customer. The next step is to remove the leg bone (femur). The thick flank (knuckle) is cut from the silverside by following the natural seam. Remove the kneecap (patella) and the internal fat deposits before further preparation of the thick flank, e.g. for diced pork or steaks.

Forefoot. The forefoot is removed by sawing through the junction between the foreshank and the forefoot bone at a right angle to the length of the foot. This foot contains some muscle and is therefore more desirable than the hind foot for food.

Fore-end. Considerable variation exists as to where the fore-end is removed. Generally one to three ribs are left on the pork fore-end. Locate the division between the third and fourth ribs from the head end and saw perpendicularly to the length of the backbone. The fore-end is trimmed of the hock which is cut off about halfway up the leg and about two-thirds of the skin and fat is removed from the butt or top end. Additionally the neckbone (all cervical and three thoracic vertebrae) and the jowl or cheek meat are removed. The jowl is removed by a straight cut parallel to the cut that separates the fore-end from the side just behind the site where the ear was removed. The fore-end may be divided into two cuts (spare-rib, also called blade Boston, and hand, also called arm picnic) by sawing just below the exposed lower end of the blade-bone parallel to the top of the shoulder. The spare-rib can be sliced into steaks or used as a roast. It can easily be made into a boneless cut by removing the corner of the blade-bone.

Besides this method some other ways of cutting and boning the pork foreend exist. In order to obtain boneless cuts (shoulder and neck-end) from the fore-end the following technique is recommended. Seam the shoulder carefully from the rest of the side, leaving the rind and associated fat behind. Release the under-blade steak and remove the blade-bone (scapula) and the shoulder-bone (humerus). Separate the main muscle block from the smaller group. The smaller group, after trimming the fat off, can be used for dicing. The main shoulder block should be trimmed of excessive connective tissue. It can be separated further into the blade and feather muscles and the main shoulder muscle.

These can then be sliced into a number of boneless steaks. The group of muscles on either side of the spinous processes of the neckbone and the two or three following segments of the backbone is called the neck-end. The neck-end is loosened from the backbone and after trimming off excessive rind, fat and any adhering ragged edges it can be cut into attractive steaks.

Lion. The middle or centre section of the pork side is divided into loin and belly by a straight cut from the edge of the tenderloin muscle on the ham end through a point on the front rib tight against the protruding edge of the split backbone. The fat back (skin and excess fat) is removed from the loin so that a complete fat cover about 0.5 cm thick remains. Starting along the backbone side at the shoulder end, cut and lift the fat over the curve of the loin muscles without cutting into the lean. The loin can be roasted whole, cut into smaller roasts or cut into chops. Shoulder, rib, loin and sirloin chops are made from the loin. Chops for broiling or frying should be cut 1.3–1.9 cm thick. Thicker chops may be made and a pocket cut into them for stuffing.

Belly. Separate the spare-ribs from the belly by cutting closely underneath the ribs beginning at the flank end. Prepare the bacon side from the belly by removing any thin or ragged pieces of lean. Turn the belly over and remove the lower edge with a straight cut just inside of the teat line. Trim the flank edge of the belly to square the whole piece to prepare it for curing.

LAMB CUTTING

Method

This procedure as described may also be followed for the processing of deer, goats, sheep or other animal carcasses of similar size.

Cooling

All lamb carcasses should be promptly chilled and kept at a low temperature (-2° to 2°C) until cut and utilized. Do not permit lamb carcasses to freeze within a day after slaughter or the meat may toughen. Lamb carcasses can be cut into retail cuts after they have been chilled for 24 to 48 hours.

Carcass

Lamb carcasses are generally not split into halves after dressing because they are not thick enough in any location to create cooling problems.

Begin cutting the lamb carcass by removing the thin cuts, i.e. flank, breast and foreleg. Lay the carcass on the cutting table and mark one side from the cod or udder fat in front of the hind leg to the elbow joint. After removing the thin cuts from both sides, remove the kidneys, kidney fat and diaphragm. Next the carcass is turned over and the neck removed either in thin slices to be braised or in one piece to be added to stew or to be boned and ground.

The trimmed carcass can then be separated into four primal cuts, each with different characteristics. A cut between the fifth and sixth rib removes the shoulder. Another cut between the twelfth and thirteenth (last) rib separates the rib from the loin. The loin and legs are separated just in front of the hip bones by cutting through the back where the curve of the leg muscles blends into the loin.

Legs. Split the legs through the centre of the backbone. Trim off the flank and cod or udder fat. Utilize the saw and knife to remove the backbone from the leg. The leg may be further trimmed by cutting through the knee-joint which is located about halfway between where the muscles of the shank end and the muscles of the lower leg begin. Work the knife and cut through the joint. Several sirloin chops may be cut from the loin end of the leg. Legs may either be prepared with the bone in or the bones completely removed and the leg rolled and tied.

The Lamb Carcass and its Cuts

Loin. The loin is usually split through the middle of the backbone and chops are cut perpendicularly to the backbone. Lamb chops are cut about 2.5 cm thick. Double or "English" chops are made from a loin that has not been split. Remove the fell or connective tissue covering before cooking chops.

Rib. The rib of lamb is prepared by sawing through the ribs on both sides of the backbone. The main portion of the backbone is then removed with a knife. Rib chops are easily made by cutting between the ribs. Remove the fell before cooking the chops. The breast portion may be barbecued in one piece or made into riblets by cutting between the ribs.

Shoulder. After splitting through the backbone, the shoulder may be roasted as is, made into chops, or boned and rolled into a roast. Arm chops should be made first by cutting parallel to the surface where the foreleg and breast were removed. Blade chops are made by cutting between ribs and sawing through the blade- and backbones. To prepare a boneless shoulder, first remove the ribs and backbone by cutting closely underneath the ribs, backbone and neck vertebrae. Next from the rear surface cut along the inside of the blade-bone to expose it and the armbone. Cut along the edges of the bones and remove them. Roll the meat and tie it securely with clean twine. The boneless shoulder may also be made into a pocket roast and stuffed with ground lamb or other dressing. The edges of the pocket roast are stitched together.

Shanks. Both the fore- and hind shanks when removed can be barbecued, cut into pieces for stew or boned and the meat ground.

Lean trimmings. Lean trimmings of lamb in chunks are suitable for stews or to be marinated and used for special roasts. Other lean trimmings can be ground and used as one would prepare ground veal or beef.

FOOD-BORNE MICROORGANISMS

Generally, food-borne microorganisms can be classified as either food-spoilage or food-poisoning, with each presenting unique characteristics and challenges to meat product safety and quality.

Food-spoilage Microorganisms

These organisms are responsible for detrimental quality changes in meat. The changes include discoloration, unpleasant odours, and physical alterations. The principal spoilage organisms are moldsand bacteria.

Molds usually appear dry and fuzzy and are white or green in colour. They can impart a musty flavour to meat. Common molds in meat include the genera *Cladosporium, Mucor,* and *Alternaria.* Slime molds produce a soft, creamy material on the surface of meat.

Common spoilage bacteria include *Pseudomonas, Acinetobacter,* and *Moraxella.* Under anaerobic conditions, such as in canned meats, spoilage can include souring, putrefaction, and gas production. This is a result of anaerobic decomposition of proteins by the bacteria.

Food-poisoning Microorganisms

Food-poisoning microorganisms can cause health problems by either intoxication or infection. Intoxication occurs when food-poisoning microorganisms produce a toxin that triggers sickness when ingested. Several different kinds of toxins are produced by the various microorganisms. These toxins usually affect the cells lining the intestinal wall, causing vomiting and diarrhea. Microorganisms capable of causing food-poisoning intoxication include *Clostridium perfringens* (found in temperature-abused cooked meats—i.e., meats that have not been stored, cooked, or reheated at the appropriate temperatures), *Staphylococcus aureus* (found in cured meats), and *Clostridium botulinum* (found in canned meats).

Infection occurs when an organism is ingested by the host, then grows inside the host and causes acute sickness and, in extreme cases, death. Common infectious bacteria capable of causing food poisoning in undercooked or contaminated meats are *Salmonella, Escherichia coli, Campylobacter jejuni,* and *Listeria monocytogenes.*

Prevention of Microbial Contamination

The initial microorganism load can be the most significant factor affecting the contamination of meat. If meat is never exposed to pathogenic microorganisms (those capable of causing human sickness), then there is no opportunity for food-borne illnesses to occur.

Several meat-processing plants have begun to utilize a program called the Hazard Analysis and Critical Control Point (HACCP) system to reduce pathogenic contamination.

This program identifies the steps in the conversion of livestock to human food where the product is at risk of contamination by microorganisms. Once identified, these points, known as critical control points, are examined to determine how to eliminate the risk of microbial contamination.

Preservation and Storage

Meat preservation helps to control spoilage by inhibiting the growth of microorganisms, slowing enzymatic activity, and preventing the oxidation of fatty acids that promote rancidity. There are many factors affecting the length of time meat products can be stored while maintaining product safety and quality.

The physical state of meat plays a role in the number of microorganisms that can grow on meat. For example, grinding meat increases the surface area, releases moisture and nutrients from the muscle fibres, and distributes surface microorganisms throughout the meat. Chemical properties of meat, such as pH and moisture content, affect the ability of microorganisms to grow on meat. Natural protective tissues (fat or skin) can prevent microbial contamination, dehydration, or other detrimental changes. Covering meats with paper or protective plastic films prevents excessive moisture loss and microbial contamination.

Cold Storage

Temperature is the most important factor influencing bacterial growth. Pathogenic bacteria do not grow well in temperatures under 3 °C (38 °F). Therefore, meat should be stored at temperatures that are as cold as possible. Refrigerated storage is the most common method of meat preservation. The typical refrigerated storage life for fresh meats is 5 to 7 days.

Freezer storage is an excellent method of meat preservation. It is important to wrap frozen meats closely in packaging that limits air contact with the meat in order to prevent moisture loss during storage.

The length of time meats are held at frozen storage also determines product quality. Under typical freezer storage of "18 °C (0 °F) beef can be stored for 6 to 12 months, lamb for 6 to 9 months, pork for 6 months, and sausage products for 2 months.

Freezing

The rate of freezing is very important in maintaining meat quality. Rapid freezing is superior; if meats are frozen slowly, large ice crystals form in the meat and rupture cell membranes. When this meat is thawed, much of the original moisture found in the meat is lost as purge (juices that flow from the meat). For this reason cryogenic freezing (the use of supercold substances such as liquid nitrogen) or other rapid methods of freezing meats are used at the commercial level to maintain maximal product quality. It is important to note, however, that freezing does not kill most microorganisms; they simply become dormant. When the meat is thawed, the spoilage continues where it left off.

Thawing meats often can cause more detrimental quality changes than freezing. In contrast to freezing,thawing should be a slow process. Meats are best thawed in the refrigerator with packaging left intact, so that moisture loss is minimized. Placing frozen meats out on a warm countertop or under warm water subjects the meat's outer layers to room temperatures for long periods of time before the meat is ready for cooking (completely thawed). This rapid method provides a conducive environment for the growth of food-borne microorganisms and increases the risk of food poisoning.

Vacuum Packaging

Oxygen is required for many bacteria to grow. For this reason most meats are vacuum-packaged, which extends the storage life under refrigerated conditions to approximately 100 days. In addition, vacuum packaging minimizes the oxidation of unsaturated fatty acids and slows the development of rancid meat.

Canning

The second most common method of meat preservation is canning. Canning involves sealing meat in a container and then heating it to destroy all microorganisms capable of food spoilage. Under normal conditions canned products can safely be stored at room temperature indefinitely. However, certain quality concerns can compel processors or vendors to recommend an optimal "sell by" date.

Drying

Drying is another common method of meat preservation. Drying removes moisture from meat products so that microorganisms cannot grow. Dry sausages, freeze-dried meats, and jerky products are all examples of dried meats capable of being stored at room temperature without rapid spoilage.

Fermentation

One ancient form of food preservation used in the meat industry is fermentation. Fermentation involves the addition of certain harmless bacteria to meat. These fermenting bacteria produce acid as they grow, lowering the pH of the meat and inhibiting the growth of many pathogenic microorganisms.

Irradiation

Irradiation, or radurization, is a pasteurization method accomplished by exposing meat to doses of radiation. Radurization is as effective as heat pasteurization in killing food-spoilage microorganisms. Irradiation of meat is accomplished by exposing meat to high-energy ionizing radiation produced either by electron accelerators or by exposure to gamma-radiation-emitting substances such as cobalt-60 or cesium-137. Irradiated products are virtually identical in character to nonirradiated products, but they have significantly lower microbial contamination. Irradiated fresh meat products still require refrigeration and packaging to prevent spoilage, but the refrigerated storage life of these products is greatly extended.

Curing and Smoking

Meat curing and smoking are two of the oldest methods of meat preservation. They not only improve the safety and shelf life of meat products but also enhance the colour and flavour. Smoking of meat decreases the available moisture on the surface of meat products, preventing microbial growth and spoilage. Meat curing, as commonly performed in products such as ham or sausage, involves the addition of mixtures containing salt, nitrite, and other preservatives.

Salt decreases the moisture in meats available to spoilage microorganisms. Nitrite prevents microorganisms from growing and retards rancidity in meats. Nitrite also produces the pink colour associated with cured products by binding (as nitric oxide) to myoglobin. However, the use of nitrite in meat products is controversial owing to its potential cancer-causing activity.

Sodium erythorbate or ascorbate is another common curing additive. It not only decreases the risks associated with the use of nitrite but also improves cured meat colour development. Other common additives include alkaline phosphates, which improve the juiciness of meat products by increasing their water-holding ability.

PROCESSING A PIG FOR MEAT

For home use purposes, you will probably want to select a young gilt (female pig) or a castrated male that weighs from 180 to 250 lbs. A smaller pig is not as economical, and a much larger pig will have a higher percentage of body weight in fat and be less manageable for home slaughter. I have slaughtered animals up to 300 lbs with good results, but the returns from feed do diminish when they get to this stage.

Judging livestock is fairly complex, can take years to learn to do properly, but here are some of the basics you can start with. Look for an animal that is broad in the back, more flat and firm than rounded, as this indicates less waste backfat. If you could balance a tea-tray on the pig's back without it wobbling, this is a desirable animal. A good wide stance is another hint that there is some solid muscle present; a pig that stands with its hind legs too close together is a less desirable animal.

The shape of the ham, the curve of the buttock and the hind leg, is something else to look at closely. A good, full, round curve means plenty of meat on the animal. A lean, angular, slanted jut to the leg and butt means that the ham is less muscular and lower yielding.

Before slaughter, pigs should be given free access to water but be denied food for at least 12 and preferably 24 hours in order to help purge the digestive system. They should not be crowded, exposed to high temperatures or unnecessarily stressed. Just before slaughter, they can be hosed clean, or scrubbed by hand if they will tolerate handling without becoming stressed. It is not only cruel to stress animals and to cause them unnecessary pain, it results in a poorer meat product.

Preparing your Workstations

You will need (ideally) three workstations and a slaughter area. You should have all these set up before the animal is hosed clean and transported to the slaughter area. An instructional manual that includes some good visual diagrams is a very good idea to have available at your first butchering. Guides to raising and harvesting home livestock are available at farm supply stores and some large bookstores, and if you do your own slaughter, one of these is essential for the diagrams on accurate and human stunning and bleeding. If you begin with the carcass, the finest reference manual for processing that I can think of is The Art of Charcuterie by Jane Grigson (listed in the bibliography).

The Slaughter Area

The slaughter area should have thick, durable plastic sheeting on the floor. This is purchasable in any hardware store and is most often found in the painter's supply aisles. Your other workstations will also need plastic sheeting, but you can get away with a much thinner and cheaper grade for those. You may use up to five 10' X 10' plastic sheets in one slaughter operation, and all but one of them can be an inexpensive housepainter's grade. If you spend $5 to $8 on your plastic sheeting, you will save yourself at least an hour and probably more of labor scrubbing surfaces, and also decrease your risk of tainted meat.

In your slaughtering area you will need a pulley or chainfall, sturdy rope of at least 500 lb test strength, a rifle or a professional livestock stunner (captive-bolt or electrical), a knife with at least an 8" to 10" blade, Rubber gloves are optional. To preserve the blood for later uses in blood pudding and blood sausage, you will need a clean bucket ready with 1/2 cup of good white vinegar already in it ready at the slaughtering station.

Workstation

Your first workstation is the dehairing area, and for this you will need a tub or 55-gallon drum full of hot water at precisely 150 to 160 degrees Fahrenheit, an instant read cook's thermometer to check the water, some large kettles of boiling water, a large sturdy table at a comfortable work height covered with plastic sheeting, a small blowtorch with a flame spreader (hardware store supply), several bell scrapers (paint scrapers will do, some long handled brushes, and so will a dull knife) and some iron or copper wool.

I have successfully used an old, heavy door on top of two sawhorses for the scraping area, covered with a sturdy and clean plastic sheet. For the chilling stage, if you do not have a cold room or very cold weather (between 32F and 40F) in which to hang a carcass to chill and firm overnight, you will need from 30 to 40 lbs of ice.

The dehairing tub should be positioned in such a way that you can dip the pig into it fairly easily by manipulating the rope and pulley or chainfall.

The tub should be sturdy enough so that the pig cannot knock it over if any unwanted shenanigans occur.

Most home hot water heaters can be made to directly deliver water at a fairly high temperature, so if you are lucky and have one in an accessible area, you may be able to get water close to this temperature out of your taps. Adjust temperature upwards by using large pots full of boiling water. In an outdoor slaughter, it is also possible to prop up a tub or drum over heavy rocks and build a fire underneath, but this may lead to overheating or uneven heating.

The sturdy dehairing table can also serves as the primal separation and processing area. Use a clean (or well hosed) plastic sheet on the scraping table. You will need many assorted knives, a knife sharpening tool, a cleaver and hammer, a meat saw, a bandsaw or a reciprocating saw, a hose with running water or several basins of water, clean paper towels, plenty of garbage bags, large food storage bags and clean food grade buckets. A clean chainsaw lubricated with food grade vegetable oil may also be used to make the backbone and heavy bone cuts on the carcass.

Brining

If you have enough cold storage room (36F to 40F) to set up overnight brine tubs, and this is highly recommended to improve the flavor and moisture retention of pork you do not plan to freeze for more than a month, you should plan for enough food grade buckets to hold as many of the following cuts as you think you want to brine: the head and trotters (for headcheese), all four legs and the loin or "pork chops".

Gutting Station

Your second workstation, which can actually be in the same place as the slaughtering area since the pig will be hanging head-down for this operation, is the gutting station. You will need access to running water or basins of water, several buckets, assorted plastic bags and meat containers, a gambrel hook or a clean, sturdy 1" X 1' dowel with a screw eye in the middle and a few feet of twine, scissors, one small (3" - 6") knife, plenty of clean paper towels and a bucket containing a 10% bleach solution or an antibacterial soap that is rated for use in food service.

Rubber or latex gloves are highly recommended. Sanitary separation must occur between this station and the later meat processing areas.

Your Kitchen

The third and final workstation is your kitchen, where you will finish the processing and preserving. Your sink and counters should be scrubbed clean with bleach before and after, and you will want to have ready a number of large Ziplock bags, many Tupperware containers, assorted knives and a knife sharpener, several good cutting boards, five pounds each of non iodized salt and loose brown sugar, plenty of coarsely ground black pepper, some spices to taste, a good quantity of fresh or dried sage, the ice left over from pig chilling in the sink, and (optional) a gallon or so of maple syrup and a bottle of good brandy or Calvados for bacon curing.

The ergonomics of the workstations should flow like this: the pig is humanely stunned while on the ground, and hauled up on the pulley or chainfall for slaughtering. The pulley must then swing over to the dehairing station as the pig is next dipped into the tub of heated water for a few minutes, then pulled out and laid on the table for scraping, and then possibly plunged in again to repeat the process. It can then be swung back to the original slaughtering area for gutting, with the original over-hock tie on its hind legs being replaced by a gambrel stick or a thick dowel threaded behind the large tendons on the hind legs and tied in place with twine.

After it is gutted and the carcass and processing area is cleaned thoroughly, the pig is allowed to chill for a few hours on ice (ideally overnight hanging in a cold room, but not all of us have a cold room) before it is swung back to lie on the cleaned scraping table, where it is separated into primal cuts and smaller cuts. Finally, those cuts are carried into the kitchen for processing into bacon, sausage and manageable cuts for the home freezer and refrigerator.

Prepare all your workstations and make sure they are clean and have all the equipment you will need ready at hand before you proceed to the slaughter.

Slaughtering

Tie a secure knot between hock and hoof on one or both of the live pig's hind legs. On an animal prone to stress, you may wish to tie a noose-type knot and slip it on just one leg as rapidly and quietly as possible. The other end of the rope should already be attached to the pulley or chainfall. Do NOT haul a live, conscious pig up in the air this way. Stun or kill the animal first before hauling it up for the bleeding process.

Shoot the pig in the forehead with your rifle at point blank range (one inch away), or use the captive bolt or electrical stunner in the same place. Draw an imaginary "X" between its ears and its eyes, and the center of the X is roughly where its brain is located. DANGER: Do not press the muzzle of a firearm directly against the pig's forehead, as the gun may jam and backfire.

Assuming the shot was on target, haul the pig into the air immediately to get a good bleed-out, high enough so that its forelegs are barely touching the ground. It will still be twitching due to the cortical disturbance triggering muscle spasm, but it is likely to be dead if a rifle was used, or effectively unconscious if a stunner was used.

If you are saving the blood, and I highly recommend it for use as garden fertilizer even if you are not a fan of blood puddings, place the clean bucket with 1/2 cup of good white vinegar between the pig's forelegs on the ground - the forelegs should be barely touching the ground. Take care not to cut the windpipe when collecting blood, or you may contaminate the bucket with weasand contents.

Stick the pig. Use a smooth dipping motion to insert the 8 to 10 inch VERY sharp preferably double bladed knife into the correct place on the pig's chest. At this point, it is advisable to find a picture with a diagram as to where the right place is, because in this instance a picture is worth a thousand words. You are aiming more or less for the heart, but it is the main arteries you will actually be severing. The motion you are making is like drawing the outlines of a shallow bowl, dipping down and then up again. If you do not get a very copious blood flow immediately, do it again quickly, with the full length of the knife.

Allow the pig to finish bleeding out, which can take several minutes. Expect some twitching and continued motion; this will persist whether or not the animal is actually dead. When you have finished bleeding out the pig, you should have 4-5 quarts of blood in your food grade bucket. Add 1/2 cup of non-iodized salt (sea salt is best, or ordinary non iodized table salt) and stir for about 30 seconds. Use a strainer to scoop off and discard the white foam. You now have stable blood that will not clot and will keep in your fridge for weeks in edible form. If you don't do this, you will have a solid block of blood jello which has much more limited culinary use. Make sure you have at least 4 quarts of blood in that bucket, or your pig is not sufficiently bled out. A good bleed-out involves a very healthy and constant stream of blood, which you can encourage some by keeping the pig's head slightly lifted so that the fat around the wound does not settle and close it.

SCRAPING AND HANGING

Dehairing or Skinning, Gambrel Hanging

Haul the bled-out pig into the tub of hot (150-160 F) water, and use the thermometer and some kettles full of boiling water to make sure it stays that way. You may want to use a long-handled brush to agitate the bristles while the pig is in the hot water. After 5 minutes or so, haul the pig out onto the table and scrape the bristles hard and furiously with your scraping tools - a bell scraper, steel wool and medium-dull knives being good candidates. If the bristles do not come off easily, dip the pig again.

IMPORTANT: Do not allow the temperature to exceed 160F, as you do not wish to cook the pig at this stage, or to set the hairs so that they cannot be scraped loose.

For the final dehairing, use the blowtorch briefly (not enough to blister the skin) and scrape the burned hair off.

If you want to skip the hot water step, you can skin the pig instead. A carcass should be thoroughly chilled before you attempt to skin, or it will be very difficult to handle. Another argument in favor of dehairing versus skinning is that pigskin is a wonderful cooking aid with which to add gelatinous body to soups and stews and to line cassoulet dishes, but before dehairing it is bristly and inedible.

As soon as the dehairing operation is over, dump out the hot water and hose the tub or drum down with cold water - you may be able to use the same large container to bury the pig in ice if you need a fast chill-down because you cannot hang the carcass in a cold room or in cold weather overnight.

Remove the rope from the hock. Carefully pierce the hind hocks behind the large tendon. Insert the gambrel hook, or thread the sturdy dowel through both hocks and tie them to the stick, widely spread apart. You may wish to notch or drill the stick to facilitate this. Attach this arrangement to the pulley or chainfall once again, so that the pig can be hung head-down for the gutting operation.

Preserving Edible Parts, Removing inedible Parts, Cleaning

Have ready a barrel, large bucket or clean garbage can lined with a big Hefty garbage sack between the pig's legs, as the word "offal" derives from a very logical etymological source - "fall off" is exactly what the internal organs do when you make that cut.

Carefully "unzip" your pig. Make a tiny cut just below the pizzle or the genital slit, being careful not to puncture any intestines, then slip your hand inside the carcass and keep two cupped fingers on the back of the knife as you cut. This keeps the guts from accidentally being slashed, Unzip the carcass very slowly and let the guts fall down unbroken out of the slit you are making. Guide them with your hands (using gloves is not a bad idea) into the bucket, and be prepared for them to really leap out of that cavity at you.

Warning: macho types who want to gut their kill with a great big Bowie knife will run into real trouble at this stage, as spilling rumen into the abdominal cavity of the carcass can taint and spoil the meat. You want a knife with a 2" to 3" blade at most for this step, or your operation will be clumsy at best and unsanitary at worst.

Pigs are the only animal I ever gut with a head-down hang, since the physical structure of the hind hocks are more suitable to take a gambrel stick and the weight of the head and neck are a strong counterbalance. If you don't mind destroying the front hocks and you want to take the head off first thing, I recommend a head-up hang so that you don't have the "leaping intestines" problem. On a head-up hang, I begin the incision just below the rib cage, which allows me to better control how the unwieldy mass of the intestines and stomach come out. You will be removing the innards in three stages: the stomach and intestinal mass in the abdominal cavity, the cluster up in the thoracic cavity behind the thin, tough membrane of the diaphragm that comprises the heart and lungs and a few other attached bits, and the kidneys and fatty bits up near the spine. The intestinal mass will be the most unwieldy and difficult, which is why the head-up hang may be the best one for beginners.

When the intestinal mass appears and begins to bulge out of the carcass, continue unzipping slowly, guiding the mass to fall down into the waiting sack. Don't forget to carefully remove the caul fat, the lacy white covering over the intestines, as this has some superior uses in cooking.

When the mass is mostly out, you will find that it is still attached to the carcass and other parts by stringy, tough connective tissue.

You will have to carefully sever any connections between the digestive system and other organs. Let the stomach fall into the barrel; it's tough and won't burst unless you were clumsy with the knife earlier. The rest of the mass will likely remain attached; fish around the diaphragm (just under the heart and lungs) with a short bladed knife that is not too sharp and find the connections to cut when you're ready to dump the stomach and guts. You may find it helpful to haul out the guts in your fists and try to have the connective tissue visible before you cut into it. Small scissors can also be invaluable at this stage.

You will come to a final length of small intestine that leads out of the body on the hind end. This is the bung. Squeeze it as clean as you can from the outside, then cut it about a foot from where it exits. Tie off both severed ends securely in a knot, and push the knot through and out the other end of the carcass.

Not far from here is the bladder, and this should likewise be removed with great care and discarded into a separate wastes bucket. Don't let go of the tube on the end of the bladder until you get over the wastes bucket.

Take a small, sharp knife and cut out the pig's entire rectum, with some flesh around it, including the tied-off bung. Carefully discard this unclean bit, without letting it touch the meat.

Do not allow any unclean material inside the lower intestines or in the bladder to touch the meat; hose off any contamination immediately. When you are done with this part of the operation and have isolated the stomach and guts into one bucket and the rumen, waste and bladder into another bucket, wash your hands in antibacterial solution or bleach water (10% solution) and change gloves before you deal with the large, edible internal organs such as the heart, lungs and liver.

If you plan to use the intestines for sausage casings (hard work but rewarding) or the stomach for Irish style haggis, you may want to begin soaking them in salt water or a light (10%) vinegar solution. The bladder may also be emptied and saved, though it has little culinary use except as a convenient container that was traditionally used for rendered lard. Sterilized glass jars work well for such storage, so the bladder can be discarded in good conscience unless you are a traditionalist or short of jars.

Beyond the thin, tough membrane of the diaphragm, fish around and grab a tough bundle of flesh up past the heart that is attaching the rest of the more solid innards to the carcass. Cut it as high up inside as you can reach, and pull. The whole mess will come down, so have another clean sack ready. This mess, except the green bubble attached to the liver, is good eating - don't waste it. Wash it well and save it on ice. You can eat the heart, the liver, the lungs, the spleen and the diaphragm, though I recommend throwing the latter scrap of tough flesh into the stock pot with theheadcheese. Remove the nasty green gallbladder from the liver carefully and discard it without letting it burst or contaminate the meat.

Fish through and separate the guts for immediate discard, immediate cold storage and short term cold storage for processing within the hour. I discard only the lower intestine, the colon and all stomach contents, and I save the stomach and upper intestines for washing and scraping clean for sausage casings and haggis, leaving them in a bucket that goes in the cold room for processing later.

If you want to process them now, here's how. Wash the stomach in running water and scrub with a brush. Store it wet, soaked in a 10% vinegar solution in cold water. Cut the intestines into 18" to 3" lengths, depending on how long you want the sausage strings. The intestines must be scraped with a slightly blunt edge - the back of a knife, a pastry scraper or a dough scraper.

Squeeze and scrape all the extraneous material out of the intestines until you are left with a translucent paperlike shell. Use more force than you think you need to. Small rips are not a major problem, and even if a length of intestine leaks water, it will often still hold sausage meat adequately.

You may wish to do some additional cleaning by attaching one end to the kitchen faucet and running water through it vigorously. Store the finished, translucent, paperlike tiny scraped intestines slightly damp, buried in noniodized salt.

Wash your hands with soap and water and a 10% bleach solution when you are done with the intestinal mass, then move on to a separate clean butchering area.

Make a moderately deep slit in both the kidneys and the top of the heart, and make sure the liver is clean of the green gallbladder. Remove the spleen (a long, dark colored strip of meat) and the spongy lungs. Place all of these organs in a food grade bucket filled with lightly salted distilled (clean) water and some ice. You want to get all blood clots out of the heart, and open up the kidneys a bit so that some of the disagreeable smell and taste can drain. Some people split the kidneys and soak them in cold milk overnight, which gives good results. Save all of these organs; they are good food.

Take a hose to the inside of the carcass once it is gutted out, or if you are field butchering away from a water source, wipe down with a damp cloth thoroughly. At this stage, it is important to allow the carcass to chill and firm, otherwise the meat will be mushy-textured, hot, slippery and hard to work with. Standard practice is to chill in a cold room or out in cold weather at 32 to 40F overnight, but I have cheated by using an ice tub and burying the carcass under about 40 lbs of ice for a few hours before continuing to work. If you do hang outdoors or in a cold room that is not sealed, you will need to shroud the meat against flies, wrapping it tightly in cheesecloth.

As soon as the meat is no longer soft, soapy and slippery, and has stiffened up a bit - it will do this fastest if hung where it is coldest, or iced down - you can commence you cutting as you wish. I advise doing the primal cuts, then hauling each of those cuts inside where you can fiddle with your books and diagrams and think carefully about how to chop them up, bone them, etc.

Tools for Separation, Disjointing, Primal Cuts

Rough diagram for primal muscle cuts:

1. Jowl bacon, a valuable cut.
2. Neck chops, less tender than the loin but still usable, located at top of shoulder
3. Shoulder. The lower section with the legbone is picnic shoulder.
4. Loin chops, the most valuable section. The rear loin will have the psoas muscle (tenderloin) included under the backbone.
5. Bacon or sowbelly.
6. Hams
7. Fatback, the solid slab of hard fat across the back of the pig

Your processing area should be very clean and well lit and stocked with many many sharp knives, a hammer and cleaver and some type of meat saw.

You can actually use the same table as you did for dehairing, but it's time to apply the hose and a new clean plastic sheet.

Remove the head by cutting deeply all the way around it and feeling with the point of your knife for the severing point between skull and vertebrae. If you fish around in here long enough to get impatient, you may wish to use the cleaver and hammer technique. Place a large cleaver blade down on the bone, and firmly strike the back of the cleaver with a hammer. Repeat until the bone is split through, and wipe or rinse away any resultant bone chips.

Remove the valuable jowl bacon squares (the cheeks and jowls) and reserve. The remainder of the head can be used for headcheese, and boiled whole or split in half after removing the eyes (some remove eyes and brain). The ears are an especial delicacy, and after being boiled into tenderness, are delicious when dipped in beaten egg and cracker crumbs and broiled with some mustard, French country style.

If you have room in your cold storage, the head will benefit from an overnight stay in the brine crock and can be placed there immediately.

After removing the head for easier handling, the first recommended cut is to split the pig down the backbone. For this cut, I recommend using a reciprocating (mechanical) saw, or a hand-held meat saw. Again, wipe or wash away the resulting bone chips. It is possible to use a cleaver and hammer to go through the backbone, but more difficult given the length of the cut and the difficulty in positioning the cleaver.

Preferably using a photo diagram, separate the shoulder and foreleg from the neck and backbone. Set this section aside for later processing in your kitchen; you can decide whether you want a whole foreleg ham, or several picnic ham and shoulder sections.

You can generally see where the joints are in the legs, and an experienced home butcher can disjoint a carcass by using a small, sharp knife to effectively take it apart at the seams. At least attempt this method, but if it fails you at first, a cleaver and hammer positioned where you think that joint is works well also.

Mark out and remove the side of bacon, slicing it off the ribs in a square that starts behind the foreleg and well under the thick, round loin muscle and goes to just in front of the hindleg. How much meat you leave on the ribs depends on whether you like spare ribs or bacon best. Skin the slab of bacon, and if you have a female pig, carefully trim and remove the mammary seeds that will be located in the flesh above the nipples.

Using a meat saw or reciprocating saw remove the rib bones just below the loin (pork chop) muscle you see attached to the backbone. A cleaver and hammer can be used, but you risk splintering the ribs if you are not already skilled with this technique. Section them as you like for storage or immediate use on the barbecue.

Working carefully with your visual diagram, remove the back leg at the hip, fleshing it away from the pelvis bone. If you plan to skin your ham, or section it into ham steaks, you may wish to skin the hind leg while it is still attached to gain better leverage. Set aside the leg for more processing in your kitchen. If you plan to skin it and trim fat away, don't put it in the brine crock yet.

The hind hocks (the ones you pierced with a dowel or gambrel stick) are often considered inedible commercially because they have been broached, but for home use, if you have used a clean dowel or gambrel, they should be fine. Separate them at the "ankle" joint and reserve them for use. You may wish to parboil them briefly in order to remove the "hoof cap", which should pry off fairly easily with a knife or a curved fork. All of the trotters (forelegs) can go in the brine crock immediately.

Take a look at what you have left. Carefully remove the skin in a thin layer, leaving a generous covering of fat - you can trim this as carefully as you like in your kitchen later on, and you probably want the pigskin pretty clean of fat for most culinary uses. Save the pigskin; it freezes nicely, and it makes a beautiful lining for a slow-cooked cassoulet as well as gelatinizing stocks and stews.

Separate the pelvis, and trim off whatever doesn't look like a nice, neat rack of pork chops. For ease of handling, I recommend cutting what's left into three sections - the neck chops, the central loin and the back loin. Look carefully at the lay of the muscle; the divisions should be obvious to you. If they are not, refer back to one of the recommended books and photo diagrams that will help you get the job done.

Shoulders, Hams, Pork Chops, Ribs, Neck, Bones

At this stage, you have a lot of large hunks of pork - two foreleg-shoulder chunks, two hindleg-ham chunks, three "pork chop" sections, two slabs each of spareribs and bacon, not to mention some scraps, the head and some internal organs. At least, that's one possible configuration - if your first attempt does not resemble anything you have ever seen in neat rows in the supermarket, take it philosophically and continue processing. The directions that follow are not particularly dependent on the shapes you cut the meat in.

What you need to know before you start cutting:

For long term storage (freezing) cut as little as possible and remove as little outer covering (fat and skin) as possible, as it will help preserve the meat. If you plan to roast immediately, only go to the work of skinning if you are health conscious and wish to also remove most of the fat. Crispy pig skin is delicious when scored into a diamond pattern by slashing with a knife and rubbed with salt before roasting.

Most cuts of pork can be improved by a day in the brine bucket - it firms up the meat considerably, and adds better moisture retention during cooking. Do not salt or even lightly brine any meat you plan to store more than one month in the freezer, as salted fat has a much shorter freezer life. If you have room in cold storage for several large brine buckets, all four legs, the loin cuts (the pork chops) and the head and trotters can benefit, *after* being trimmed of fat to the desired degree, from an overnight soak in brine. Do not brine any pieces before trimming the fat to the desired degree.

Pieces meant for brine cure should be skinned and a knitting needle inserted and removed alongside the bone at both ends, several times. Pieces meant for cold cure should be skinned and deboned. Full penetration of brine and cold cure is critical for a healthy and successful preservation. This is not necessary for the overnight brine cure, only for the pieces destined for bacon or ham.

Begin with many, many sharp knives and have a helper who will continually resharpen them. You will dull knives faster than you think while you are doing this, especially for the first time. Good quality sharp knives will save you an amazing amount of time, cussing and hurt.

Have ready a lot of heavy duty freezer bags and an indelible marker to record what is in them, or you will be scratching your head at all the mystery bags in the freezer for months to come.

At this point, you'll have to decide what cuts will be right for you and your household. If you plan to throw some major parties soon, you may leave some of the larger cuts intact (legs, whole loin roast) as an impressive menu item for the feast. If you have limited storage space, you may want to debone and throw most of the shoulders into the cure with the bacon (yes, this does produce an excellent product in the cold cure), make a lot of terrine and sausage, debone the legs into three or four roasts and remove the loin muscles (both on top and underneath the backbone) for easier storage.

Generally, a good and simple plan for the average household goes like this:

Forelegs: Take the unskinned picnic shoulder from the foreleg, separating it by the lay of the muscle from the meat attached to the shoulderblade. You should have a nice, round small ham. This can be left unskinned and frozen or roasted immediately, skinned and cured in brine, or skinned and boned and cured in dry cure.

Remove and trim the neck areas from the shoulder roast; use trim in headcheese,terrine or sausage. The shoulder roast may be separated into two pieces by sawing through the shoulderblade, or you can ask your butcher to bandsaw the shoulder into steaks for immediate use. You can also bone out the meat if storage space is a concern, or slab it into 1 1/4" thick squares or rectangles for curing along with the bacon. Shoulder meat is ideal sausage and terrine meat, and can also be preserved in rillettes or rillons.

Neck: You may bone out the neck and use the meat for headcheese, terrine or sausage. Some people hot-smoke the neckbones, or barbecue them.

Head, hocks and tail: Put them in the pot of boiling water or the brine for headcheese, along with any other random trim that looks gristly. Even if you plan to pickle the feet in vinegar and brine later, they benefit from parboiling.

Spare ribs: Cut into manageable sections for storage, or better yet, barbecue them immediately as they take up a lot of room you will be needing for the meatier parts. This also avoids the annoying problem of cutting them if you don't have a good butcher's bandsaw.

Hind legs: You have a classic "ham" here, and you may wish to brine cure and smoke these pieces, or dissect into much smaller pieces for pressed dry cure, immediate use or freezing. An excellent cold cure can be achieved on a deboned ham using the same recipe given for bacon, though the cure takes a few days longer for the thicker pieces, and is done when the meat is uniformly firmed up to the touch. Ham definitely benefits from some time in the brine bucket if it's convenient. If you do not debone the leg, insert a knitting needle down into the meat next to the bone on several sides to help facilitate the brine penetration.

Pork chops: You should have three sections of "chops", one less desirable section from the neck which is often boned out for sausage meat or headcheese, one middle section and one back loin with a tenderloin "eye". Cut these sections into manageable parts if you can't easily store them in these lengths. Slice through the meat neatly with a sharp knife, then use the cleaver and hammer to go through the backbone.

If you plan to use them within one month, definitely give them one day in the brine crock to improve their consistency and flavor. Trim the fat covering to your liking before they go into the brine, but don't leave them naked or they will cook up dry and tough. If storage space is a serious issue, debone the chops and keep only the tenderloin, processing the bones immediately for stock.

Bones: Save the bones. They make excellent stock. Roast them in the oven at 350 degrees for half an hour with whole onions, tomatoes and carrots, then boil the lot with some bay leaves, celery, salt and pepper to taste. Pork stock is rich and delicious, and can be used as the base for many soups, stews and bean dishes.

Preserving and Rendering

Leaf lard or kidney fat: A thick layer of fat on the inside of the carcass, often completely enveloping the kidneys. This is extremely desirable lard, and should be rendered separately from the coarser grades.

Sowbelly or bacon: This is streaky fat with both hard and soft textured fat mixed in with muscle. It lies along the flanks of the pig, and is generally cured and eaten as bacon.

Fatback or hard fat: This can be salted or cured like bacon for keeping, rendered for good quality lard or used to line terrine pans or to tie around leaner roasts, especially game roasts, to keep them tender. You mostly find fatback (where else) on a pig's back, in great thick sheets next to the skin.

Soft or soapy fat: This is the trim, the scrapings, the gristly soft stuff in and around the muscle tissue. It isn't good for much, but you can melt it to make an inferior quality lard. Don't mix this fat in with any other kind of fat; it is not very good for sausage making either.

Caul fat: This is the lacy membrane surrounding the stomach and intestines. It makes poor lard, but it is very attractive as a covering for pates and terrines, to wrap around meat, or to make French country style gayette sausages.

Identify all the different types of fat you will be removing from the carcass, separate them, and use them appropriately.

Rendering lard is not a difficult task. You can start with soaking the fat in cold water overnight, which helps remove any blood or impurities, or you can cut the impurities out and discard them if you don't mind a little more waste. Chop or grind the fat coarsely. Place the fat (separating into different pans by type of fat) in a heavy cast iron pan with just enough water to keep it from burning, perhaps a quarter inch. Put the pan on a low, steady heat, stirring very occasionally, until melted. You can also render lard in pans in the oven at about 300 degrees.

Pour off the fat as it melts, and store it in clean, boiled glass jars. You can salt and eat the crackings left over, but don't let any salt get into your lard, or it will go rancid faster.

Headcheese

Headcheese is one of the more simple and popular uses for the pig's head, though there are certainly more complex ones - deboning and stuffing, roasting whole, or deboning cheek and jowl in the classic English "Bath Chap" dish are all methods of utilizing this underappreciated cut that have a long historical backing.

If you have enough room in the brine crock to soak the head and trotters overnight, it does improve the flavor for anything you're thinking of doing to them, including headcheese. But it's optional; some people don't have enough cold storage room for that much pig.

To make headcheese, remove the eyes from the pig's head and set it in a large pot to slowly boil down for four or five hours in hot water and a generous dollop of white vinegar. If you like, add a handful of herbs to taste (bay leaf, thyme and cracked pepper works well), some onions, carrots and garlic. Add all four cleaned hocks as well, and make sure you have neatly de-hooved them first. You can add any meat scraps that you do not want for terrine or pate or sausage also, especially if they are gristly bits that will benefit from long stewing.

You can add those trotters to the headcheese, or you can do something else with them once they have been boiled into softness. You can pickle them in a mixture of vinegar and brine, or bone them and stuff them and roll them in crispy breadcrumbs. The ears, boiled into tenderness on the outside and crunchy cartilaginousness within, are also worth removing and serving up separately.

Splitting the head in half is optional but helpful; if you do, remove the brains and reserve for another use. By the time you're done butchering at the end of the day, about 5 hours later, you'll be ready to collect up all the tasty bits that have fallen off the bone and press them into a nice brick of headcheese.

Continue reducing the liquid in the pot to a well flavored jelly, adding wine, salt, lemon juice, herbs and vinegar to your taste. Chop the meat coarsely into rough dice - sorry, no shortcuts here, as the food processor or grinder will turn your headcheese to mush. Layer the meat as interestingly as you like. I sometimes use a whole tongue in the middle, and some pistachio nuts for crunch. Cover with the reduced liquid and allow to set in the cold room or refrigerator overnight If you've reduced the liquid enough to just cover the meat, it will set into an attractive jelly. Slice and serve cold.

Making Sausage - From Scratch

Wear latex gloves for this next part of the job. If you are a complete wuss and don't even want to do this with gloves on, throw away the guts and go buy sausage casings. Or don't, if you plan to make only coarse pork sausage, which freezes very well indeed in bulk.

In an area well away from the clean meat (which should be in your fridge, in a brine crock or in your kitchen sink on a bed of ice right now), empty as much of the rumen as possible out of the stomach and intestines in a garbage bag. Tie off and throw away the garbage bag. Take the garden hose to the mess; wash it as well as you can.

Section the intestines into lengths of roughly 18" to 2 feet. You can keep them longer if you want, but the scraping gets problematical after awhile.

Again, squeeze clean, soak and wash. Use a wooden chopstick to turn the intestine lengths inside out, wash some more. You can also run water directly through the lengths of gut, by stretching them around your kitchen sink faucet.

If you prefer chittlins or prepared pork intestines to sausage casings, by all means reserve some or all at this stage, before scraping. Laying each intestine section on a sturdy plastic cutting board that you can adequately sterilize afterward, scrape it firmly and repeatedly with a dull knife or blunt edged scraper until it is a translucent casing. Practice makes perfect. Don't mind the small holes; they happen, and they will not interfere with your sausage making. Discard all of the slimy white material that comes off of the casing; this is the intestine lining.

Do a final salt water wash on the casings, dry them briefly and pack them in salt. They'll keep for weeks that way, until you are ready to make your sausages.

There are innumerable recipes for sausage, but a good basic one begins with the proportions. Standard proportions are 1 lb lean pork (neck or shoulder) to 1/3 to 1/2 pound fatback or sowbelly, 2 teaspoons of salt and spices to taste. Coarsely grind or run through the food processor on slow or pulse speed, being careful not to reduce the meat to mush. Be sure the meat and hard fat is chilled thoroughly before you begin grinding.

Some recipes double the amount of fat, calling for equal proportions. Since you are making your own product, there is no reason you should not experiment at your leisure and produce leaner sausages if you wish. You should be aware that much of the good taste of pork sausage does depend on fat, so if you wish a lower fat dish, you might want to use those casings to produce chicken or seafood sausages instead. Experiment and see where your tastes are.

To stuff sausage meat into your casings, you can use the extruder on a large pasta machine, buy a special attachment for your food processor, or buy an old fashioned sausage plunger. Follow any standard recipe for sausage, using your own desired proportions of fat and lean and spices, and stuff those casings you so laboriously scraped. Do not stuff them overfull. Twist them to form the sausages on the string, and you can crimp them off as you go, or tie knots and make individual sausages.

The more fat contained in your sausages, the more successfully they will keep in your freezer. Glazing sausages with additional melted lard to protect them in the freezer is also a good idea.

Casings are optional for everything but the softer types of sausage, such as blood sausages. Pork sausage patties or balls freeze very well in Ziplock freezer bags. Crepinettes or gayettes substitute caul fat for casing, with a round ball of sausage wrapped in a square of caul fat which has been soaked in warm water and dried off for pliability.

Another excellent sausage "casing" is the skin from a goose, duck or turkey neck. You can fill this with a mixture of sausage meat, chestnuts, rice or even truffles if you are fortunate enough to possess any, and roast it crisp.

Makin' Bacon

Trim your skinned sides of bacon, removing as much fat as you wish, and taking care to remove any mammary seeds (small, hard granules of tissue) from the tissue over the nipples if you have taken a female pig. Prepare a single flat container (I often use very large plastic storage or shoebox containers) by mixing 1 pound of brown sugar to 1 pound of non-iodized salt (sea salt works well) in it. Add cracked black pepper or other seasonings to taste. Cover the sides of bacon with this mixture thoroughly, rubbing it into the meat on every side. Leave no meat surface untouched. Use more of the mixture if necessary; the proportions are fairly simple.

If you wish to cut up the bacon in smaller squares for easier stacking during the curing process in your fridge, this is fine. I tend to add other cuts of meat to the bacon cure, such as fatty shoulder meat and other tempting looking bits off the pig that I think would do well as cured meat. It's all fair game if you slab it off into pieces no more than an inch and a quarter thick to take the cure.

Stack the slabs of bacon, making sure there is a generous thick layer of the salt-sugar mixture in between them, under them and on top of them. Cover with plastic wrap and weight down in the refrigerator or in a cold room (no more than 40 degrees F) with heavy bricks.

After 24 hours, drain all liquid from the mixture and add more of the dry cure (1/2 salt, 1/2 sugar, spices to taste). Leave it for another 24 hours, then drain again. This time add liquids - maple syrup is an excellent additive, or you can use apple brandy, or both. Other fun additives (but not all at once): bay leaves, rum, nutmeg, sage, liquid smoke flavor.

Some people recommend using a pinch of saltpeter in this stage of the cure. I don't. There isn't a lot of point in going to all this work to do your own curing if you aren't making a better product than you could find in the store, and a natural home sugar cure far surpasses what you can get with chemicals, in my opinion.

Let the bacon sit until the cure has fully penetrated, generally another day or two. Slice off pieces and test it periodically.

Home sugar cured bacon cannot be fried on high heat; the sugar will caramelize and burn. Fry it slowly on very gentle heat, and the taste sensation will reward you. You can keep successfully cold cured bacon unspoiled for many months in your refrigerator in Ziplock bags with a preserving splash of brandy.

Ham

The basic recipe is pretty simple. Leave a joint of pork in brine in cold room storage (36F to 40F) for a minimum of three days (for a small, skinned and boned cut) and a maximum of 30 days. You can also use a dry cure after the first day of brining, with the same recipe as for bacon - 1/2 noniodized salt, 1/2 brown sugar, spices to taste. Press it under bricks overnight, and when the liquid runs off, you can drain it (or keep it if you're thrifty) and add enough maple syrup and apple brandy to cover it up again. The meat is done when slightly firmed to the touch. An overcured ham will be salty and will need long soaking, or several changes of boiling water, or both.

When you're ready to cook it, parboil the ham for about 10 minutes and taste the water - if unpleasantly salty, discard the water and parboil again. When it stays clean and sweet, add some carrots, onions, celery, bay leaves and pepper and simmer slowly for about 30 minutes per pound.

For the final glaze, crust it with mustard, brown sugar and pineapple juice (or any other recipe that suits you) and bake it at 325 for 20 to 30 minutes, making sure the sugar does not burn. Basic brine recipes and additional ham curing tips are in Section X. Warning: do not attempt to cold cure any meats outside of your refrigerator unless you really know what you are doing and/or have a cold room or reliably cold weather in your area, from 36 to 40F.

Variety Meats

The heart, kidney and liver will benefit from overnight soaking in milk or salt water to help remove blood clots and any strong or objectionable flavors. The spleen is edible as it is, though it is somewhat tough.

Lungs are very, very tasty, and for no particularly good reason beyond cultural food taboos illegal to sell in America. Beat them flat with a heavy rolling pin or cast iron frying pan and boil them for a few hours, changing water as needed if the smell is not appetizing yet. Eat them or add them to a recipe when the smell is appetizing and the texture very tender. Lungs need long stewing, but my they are utterly delicious when you're done. You may wish to remove some of the more cartilaginous and gristly bits first. This can be done either before or after cooking.

Pig stomachs need to be cleaned and scrubbed thoroughly, and boiled for at least an hour before they begin to be palatable. They can be served in Chinese style recipes as tripe, or stuffed with stewed kidney, heart, liver, spleen and lungs along with some mashed potato to make Irish haggis.

Fresh pork liver makes a beautiful terrine or classic French pate, and can also be used in most recipes calling for beef liver. It is tougher than beef liver, and is generally at its best when ground or chopped.

The caul fat, the lacy white "net" that you removed from the intestines and soaked, is excellent for barding roasts or for making a classic French dish, gayettes. You can also line a terrine or pate dish with the caul fat, and it makes a very attractive alternative to bacon or fatback.

Long term Storage

There are several ways to achieve good long term storage of your pork. You can use brining or dry curing with salt, or make rillettes or confit, where pork that has been salted overnight and then wiped clean is slowly cooked in its own fat until almost all of the water in the meat has been replaced by fat and salt. Rillettes can should keep for many months, covered completely with fat, in a cool and dark place.

For every pound of meat, use half a pound or more of good lard or unrendered fat. Use sowbelly or shoulder for this long-lasting French country delicacy. Cut your rilletes (cubes) into 2-inch dice. Cut your rillons (shredded pork) into thinner ribbons of meat, about an inch long and 1/4 inch wide. You can make either or both, and in the same pan if you like.

Place the meat and fat in a heavy cast iron pan, season with salt and pepper and spices to taste, and cook covered at about 300 degrees for 4 hours. Separate the rillettes (cubes) and brown them more crisply in another pan, uncovered. If you plan to preserve them for some time, cook a great deal of the moisture out of them. If you plan to keep them refrigerated and eat them within a week, you can leave them moist. Make sure all of the meat juices end up evaporated or separated from the fat.

Strain the remaining shreds of meat out from the fat. Using a wooden mallet, or a wide meshed sieve, pound and rub and shred the meat into fibers. A food processor or grinder does not suffice for this task, and it will reduce the tasty fibers to unappetizing mush. Season with spices to taste.

Pack the meat (both rillons and rillettes, together or separately) into sterilized glass jars. Pour the clean melted fat on top of them. Do not allow any liquid (water, meat juices) to enter. If the meat is completely covered with lard to a depth of at least half an inch, it will keep in a cool, dry place for months. Cover the jars with cheesecloth, aluminum foil or standard lids, after the lard has cooled and hardened.

Smoking and Drying

Smoking and dry air curing (prosciutto) are two other techniques that have been used historically for long term preservation. Hot smoking is easier than cold smoking, and both are easier and less risky than dry air curing.

Cold smoking adds additional flavor and some preservation value to meat that has already spent a few days in brine. It is best done with plenty of air flow and a temperature of an even 60F. Unless your technique is very meticulous, there is some risk of contamination, so hot smoking is better recommended to the amateur.

One technique of semi-cold smoking mentioned by Jane Grigson is to keep an intense flow of smoke at a temperature below 90F for an hour at a time, allowing the side of bacon or ham two days to dry and cool in between smokings to cool down between applications of smoke. Two or three such intensive sessions is enough to allow the smoky flavor to permeate. Do not allow the temperature to get above 90F at any time, or you will find your bacon melted and your ham toughened. The first smoking session should last half an hour, and the meat is rubbed with peppercorns, spices, thyme and crumbled bay leaves while still warm.

If you get good results, you could move on to the second stage of curing, where the ham continues to be smoked for a few hours a day and dried and cooled in between, until it loses a full quarter of its weight and turns a deep golden brown. Remember that if the process goes wrong at any time, you risk losing a ham. Do not eat or feed to animals any meat that you think went "off" in the curing process.

For long term storage of a ham that has been sufficiently cured, shroud it with boiled sturdy cloth (canvas or linen) and hang it in a cool, dry, dark place where the temperature range does not go below 32F or above 60F. The ideal storage range for fully smoked hams is 55F to 60F. Humidity is a worse enemy than temperature, as moisture will soon cause your hams to grow mold.

Fortunately, most surface molds are harmless and can be cut away without damaging the meat beneath. Hams can also be stored well wrapped in cloth or paper in boxes packed with a thick bedding of cottonseed hulls, raw oats or grain. Bury the hams in the grains, and check them periodically to make sure they have not become infested with insects.

A good place to start experimenting with hot smoking for long term preservation is with pork neck bones and hocks and other inexpensive cuts. You can also hot smoke for immediate consumption. Experiment, and follow the directions on the smoker that you purchase. A very good stovetop model is made by Cameron. Here is one recipe for an air-dry cure. Although it is supposed to be eaten raw, please consider that there is always some possibility of contamination in a nonsterile home processing operation, or disease in the animal, and proceed with your own safety in mind.

Dried Sausage, Raw Cure (traditional recipe)

Basic dried sausage recipe: 1 lb lean pork to 1/2 lb fatback, 2 tsp salt, 1/2 tsp each of pepper, sugar, spices to taste, 1 clove raw garlic, 1/2 tsp saltpeter

Grind all ingredients together finely. Fill some large casings (do not overfill), keeping in mind that this sausage will dry and shrink. Hang at 60F or below, preferably with a fan running and plenty of air circulation. As the sausage shrinks, squeeze it down so that it remains compacted. This is a raw, cured pork product, and you should bear this in mind when you consider the safety of making it at home. It is traditionally sliced and eaten raw, but you may wish to cook it for safety.

When all is said and done, the comforts and safety of modern technology are nice to have around. You will probably want to preserve a significant portion of your meat in the freezer.

Freezing

You can freeze pork for a fairly long time, though an average sized household that also entertains guests can completely consume one pig over a matter of not very many months. What I find to be most economical is sharing a butchering with at least one other household, so that I can do them more often and the meat is fresher.

Keep sections destined for the freezer as large as possible if you are looking at long term storage. Cut into smaller, convenient portions only what you plan to pull out of the freezer and use inside of a month. You may also consider investing in a saw that can cut meat while it is still frozen, making it more convenient to store meat in larger and longer lived pieces in the freezer.

To help protect against freezer burn, use melted lard or even vegetable shortening as a thick coating around the meat, and wrap that up in several layers of plastic wrap to be followed with heavy freezer plastic. Make sure the lard or shortening is pure and free of salt. Freeze pork within 8 days of slaughtering.

Pork can be stored up to 8 months in the freezer if it is not brined or salted, or 2 to 4 months if it has been brined. Ideally, it should be initially "flash frozen" at a temperature of -10F or below, if you can manage this in a chest freezer or by paying a commercial facility if you have only an upright freezer available. The recommended ratio for freezing food is no more than 2 lbs per cubic feet of freezer space for the initial freezing, well spaced out. After the meat is solidly frozen, you can set the temperature controls back to 0F.

Freezing cured or smoked meats often produces tasteless, mushy or poor results, unless they are first cooked.

EFFECT OF CONTAMINATION ON SENSORIC PROPERTIES OF MEAT AND MEAT PRODUCTS

Aerobic spoilage by bacteria and yeasts usually results in slime formation, undesirable odours and flavours (taints). Colour changes, rancidity, tallowy or chalky flavours from the breakdown of lipids may also occur. Colour changes as a result of pigment oxidation may be grey, brown or green discoloration. Aerobic spoilage by moulds results in a sticky surface, musty odours, alcohol flavours and creamy, black or green discoloration.

Anaerobic spoilage which occurs either within the meat or on the surface in sealed containers where oxygen is absent or very limited is marked by a souring due to the production of organic acids and gases.

Food Poisoning

Food poisoning may be due to infection or intoxication. Infection is caused by the consumption of live bacteria which multiply in the body producing characteristic symptoms. Intoxication is due to toxins in food produced by bacteria before the food was eaten. Toxins are chemical compounds which may linger in food with no microbes growing in it, and are therefore very dangerous. Salmonellae are facultative anaerobes which cause infectious food poisoning. Ten or 20 cells of *Salmonella typhi* are sufficient to cause typhoid but 10 000 to 100 000 cells of other species may be necessary to cause an infection. Some are host-specific affecting the animal from which the meat was produced but failing to cause infection when consumed by man. Typical symptoms of salmonellosis include diarrhoea, fever and vomiting. The illness may last one to 14 days after a 12 to 24-hour incubation period. Victims may excrete the bacteria for weeks after the symptoms subside. Poor personal hygiene will cause contamination of meat.

Staphylococcus aureus is a facultative aerobe that causes intoxication. It lives in the nose, throat, hair and skin and on animal hides. Meat is contaminated by handling and by sneezing or coughing. Minute amounts of the toxin will cause illness, which starts within one to eight hours of eating poisoned food. Nausea, vomiting and shock may last for one to two days. On rare occasions it is fatal. This bacterium does not produce off-odours or spoilage so it cannot be easily checked. Refrigeration will control its growth. Cooking may destroy the bacteria but not the toxin as it is heat stable. It is particularly troublesome in cooked cured meats, normally as a result of recontamination after the curing process in subsequent handling, for instance during slicing.

Clostridium botulinum, an anaerobe, produces the toxin botulin, one of the most poisonous substances known. This attacks the central nervous system causing death by respiratory paralysis. Dormant cells occur everywhere in the soil, fish, animals and plants. High-moisture, low-acid, low-salt conditions at above 3°C favour growth and toxin production. Control measures must destroy spores or prevent growth and toxin formation. Botulism is usually due to undercooking processed meats. Pressure-cooking will give commercial sterility. Pasteurization (heating to 70°C) and adding salt (NaCl) and sodium nitrite ($NaNO_2$) is used for canned ham. Refrigeration (0–10°C) is essential for vaccum-packed meats. Frozen storage prevents growth.

Clostridium perfringens, an anaerobic bacterium, is a common cause of food poisoning but is rarely fatal. It grows well in warm meats so is usually found in left-over meats that have not been kept chilled and not been reheated to 70°C to kill the bacteria present. The main symptoms are diarrhoea and weakness which last for 12 to 24 hours after an incubation period of eight to 20 hours.

6

Raw Meat: Cross Contamination

Cross-contamination is the transfer of harmful bacteria from one food to another - this may be directly, or via chopping board, utensils, hands, etc. Cross contamination occurs when handling raw meat, especially poultry meat, because it contains more liquid than other meats. It is important to keep meats and their juices away from already cooked or ready-to-eat foods such as salads.

Uncooked poultry meat is frequently contaminated with Campylobacterbut rarely with Salmonella. These bacteria are naturally present in the gut of birds, indeed all animals, and can transfer to the meat during processing.

Campylobacter is a major cause of foodborne gastroenteritis in New Zealand and raw meats, especially poultry, is one of the sources for its transmission. If you become infected with Campylobacter you are likely to become ill within 2-5 days with symptoms of muscle pain, headache and fever followed by diarrhoea and abdominal pain. The illness lasts about 5 days.

There is no way of telling that meat is contaminated with Campylobacteror Salmonella. Therefore, it is advisable to treat all meat as potentially contaminated.

To ensure you don't cross contaminate when handling raw meat:

- *When shopping:* Separate raw meat from other foods in your shopping trolley. Place these foods in plastic bags to prevent their juices from dripping onto other foods. The poultry industry now provides whole chickens in leak-proof packs and these should be selected wherever possible
- *In the refrigerator:* Place raw meat in containers to prevent their juices dripping onto other foods. Juices will contain harmful bacteria if they are present on the meat
- *Defrosting:* Completely defrost meat before cooking so it will cook evenly. Ensure juices do not drip onto other foods

- *In the kitchen:* Harmful bacteria can spread throughout the kitchen and get onto chopping boards, utensils, and bench tops. To prevent this:
  - Wash hands with soap and hot water before and after handling raw meat
  - Wash chopping boards, dishes, utensils and benches with hot, soapy water after preparing each food item. Preferably, use one chopping board for fresh produce and a separate one for raw meat
  - It is not necessary to rinse raw meat before cooking. Studies have shown that rinsing can spread bacteria around the kitchen
  - Cook raw meat thoroughly until juices run clear. This is especially important for poultry meat
  - When marinating, boil used marinade before brushing onto cooked meat
  - When serving, never put cooked food back on the same plate or chopping board that previously held raw meat.

Some additional cleaning tips to prevent cross contamination:

- Microwave ovens provide an effective means of sterilising wooden chopping boards (3-4 min, 800W) and cloths (wrap in paper towel and microwave until steaming, 30 sec-1 min). Rinse under tap to remove surface dirt before microwaving
- Cleaning chopping boards and cloths in a dishwasher is also an effective method of eliminating harmful bacteria
- Boards and benches can be wiped with a dilute solution of bleach (1 teaspoon bleach in 2 litres water) following cleaning to ensure harmful bacteria are destroyed.

Definition

The product components muscle meat, fat and non-meat ingredients which are processed raw, *i.e.,* uncooked by comminuting and mixing.

The resulting viscous mix/batter is portioned in sausages or otherwise and thereafter submitted to heat treatment, *i.e.* "cooked".

The heat treatment induces protein coagulation which results in a typical firm-elastic texture for raw-cooked products. In addition to the typical texture the desired palatability and a certain degree of bacterial stability is achieved.

Precooked-Cooked Meat Products

Precooked-cooked meat products contain mixes of lower-grade muscle trimmings, fatty tissues, head meat, animal feet, animal skin, blood, liver and other edible slaughter by-products. There are two heat treatment procedures involved in the manufacture of precooked-cooked products.

The first heat treatment is the precooking of raw meat materials and the second heat treatment the cooking of the finished product mix at the end of the processing stage. Precooked-cooked meat products are distinguished from the other categories of processed meat products by precooking the raw materials prior to grinding or chopping, but also by utilising the greatest variety of meat, animal by-product and non-meat ingredients.

Recognising Properly Cooked Meat

Unfortunately, colour is not a reliable indicator of properly cooked meat. In fact, meat may appear brown before it is fully cooked, especially for whole cuts of meat. On the other hand, some meat may look pink even when it is fully cooked.

Many things can affect the colour of meat in addition to whether it is cooked thoroughly or not. For instance, the amount of fat may affect the meat's colour and so can vegetables cooked with meat.

Pathogens may be outside or inside the meat, so internal temperature is important and the only way to know for certain that it has been cooked safely.

Meat Thermometers

The only way to recognise properly cooked meat is with a meat thermometer. A meat thermometer can help you:

- Prevent foodborne illness.
- Prevent overcooking.
- Allow meat to be held at a safe temperature.

There are several types of meat thermometers:

- Oven-proof
- Instant read or digital
- Pop-up
- Microwave-safe

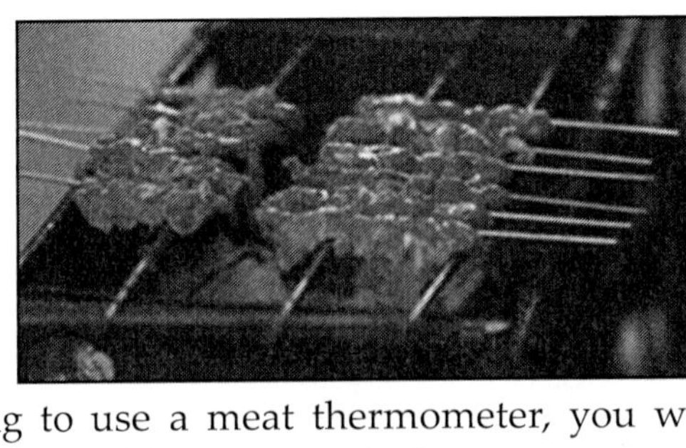

If you're going to use a meat thermometer, you want to be sure your thermometer is accurate. You can test it for accuracy easily by inserting the thermometer two inches into boiling water. At sea level, it should read 212°F. It will be 2°F lower than 212°F for every 1,000 feet above sea level. For example, water boils at 200°F in Denver, Colorado. Some thermometers can be calibrated or adjusted under the dial. Next, you want to be sure that you're placing the thermometer in the meat correctly. For poultry, insert the thermometer at the inner thigh near the breast. For ground meat, insert the thermometer at the thickest area. For whole meat, insert the thermometer in the thickest area but away from bone.

Finally, you want to insert the thermometer at the correct time. Oven-proof thermometers can be inserted at the beginning of the cooking time. Instant-read or digital thermometers should be inserted when you remove the meat from the oven. Remember to wash your thermometer after each check of temperature.

Meat can be Cooked in Some Types of Flexible Materials

Oven cooking bags are acceptable for baking, but grocery bags should not be used for cooking. Hock locks can be left on turkey and poultry during cooking, but be sure to remove any internal organs from the poultry cavity.

Plastic bone guards in ham should be removed before cooking. However, netting around ham or turkey can be cooked and pop-up temperature indicators can be left in while cooking. Remember to remove these items before serving. Be sure to remove any absorbent paper and plastic pads that are often underneath the meat in its store package.

Meat should never be cooked or reheated in a plastic container unless the container is labelled microwave safe.

Raw-fermented Sausages

Fermented Sausage Production

The broad group of fermented or raw sausages includes a large number of products whose characteristic properties are partially or completely dependent on fermentative action of certain types of bacteria. The comminuted meat mass may be submitted to curing either prior to or after stuffing. The stuffed sausages are processed by smoking, drying and ageing which make a product entirely suitable for eating without further cooking.

The principle two subgroups of fermented sausages are semidry or quickly fermented and dry or slowly fermented (semifermented) sausages. There are both hard and soft types in both subgroups.

Semidry Sausages

Semidry sausages differ greatly from dry sausages by their pronounced "tangy" flavour of forced fermentation resulting in lactic acid accumulation and a bulk of other products of fermentative breakdown. The addition of starter cultures for a number of semidry sausages is particularly successful.

Semidry sausages are usually stuffed in medium-and large-diameter natural or artificial casings. The length of production (smoking and fermentation) of these sausages depends upon their type but rarely exceeds several days.

The pH of semidry sausages is explicitly acid (4.8 to 5.2-5.4); although they are often finely chopped and spreadable, many of them can be cut in thin slices; their water content reaches 35 percent or more.

Semidry sausages are regularly smoked and only exceptionally slightly cooked by the heat applied in the smokehouse at various temperatures, mostly not exceeding 45°C and very occasionally rising to nearly 60°C for a strictly limited time; after smoking the sausages are usually air-dried for a relatively short time.

Semidry sausages usually contain an important proportion of beef. Their shelf life is surprisingly good due to low water activity, accumulation of acids and smoke compounds, counteracting the effect of lactic acid bacteria on spoilage microorganisms, etc. A high level of hygiene and the ability to perform dexterously all operations in the manufacturing process are basic prerequisites for the good keeping quality of semidry sausages. Semidry sausages have improved stability if stored in the chiller, protected from humidity rather than at room temperature. This category of sausages is popular in many European countries and North America. As these sausages need only little refrigeration, they can be successfully produced in many subtropical countries.

The main sausages of this group are: summer sausages (with a series of varieties in many countries), different types of cervelats and metwursts, lebanon bologna (in USA), etc.

Dry Sausages

The organoleptic and other properties of dry sausages depend not only upon the products of sugar bacterial fermentation but are also strongly influenced by biochemical and physical changes occurring during the long drying or ageing process. The use of starter cultures for this category of raw sausages is less successful than for the semidry varieties. The length of production, either with or without smoking, and drying periods depends upon a multiplicity of factors, such as diameter and physical properties of casings, sausage formulation, choice and methods of preparing meat, conditions of drying, etc., but overall processing time may require up to 90 days. The final pH of dry sausages is usually somewhat higher (5.0-5.5) than in semidry sausages, and it increases during the second part of this long ageing process.

Dry sausages are made from selected, mainly coarsely chopped, meat (some Italian salamis, some types of sucuk); often they are moderately chopped (majority of small-diameter dry sausages) and very occasionally finely chopped. They are cut in thin slices, their water content is under 35 percent, but normally less than 30 percent. Most varieties of dry sausages are subjected to cold smoking (12 to 18°C) but sometimes not; in some countries they are often heavily spiced with red pepper or garlic or sometimes heavily smoked and strongly salted. In principle, they are processed by long, continous air-drying, sometimes after a comparatively short period of smoking.

The formulation, degree of grinding, level of fermentation, smoking intensity, temperature of ageing and type and size of casing as well as other factors determine the properties of the final product. Dry sausages are stuffed in both natural and artificial casings of varying diameters. In the preparation of dry sausages natural casings are preferred because they adhere closely to the sausages as sausages shrink. Sausages stuffed in casings with a diameter exceeding 4.5 cm are often called "salami"; salamis are chiefly made from coarsely ground meat, predominantly of pork and are not smoked. The shelf life of dry sausages is excellent, which may be especially attributed to the high salt-to-moisture ratio. These sausages are normally kept without refrigeration.

Raw sausages, which are not submitted to the smoking process, are known as air-dried sausages. This variety of dry sausage is characterised by a highly attractive appearance and by its yeasty-cheesy flavour. Air-dried sausages are marked with or without mould overlay.

The principle dry sausages are salami of different types produced in many countries followed by cervelats and many small-diameter dry sausages. Dry sausages may be hard, intended for slicing, and soft style sausages, which can be spread. An especially important fact in dry sausage manufacture is that the same raw materials, stuffed into different casings, give a quite different finished product. The wrapping of the twine on the sausage, contributing to the strength of the casing, is also of importance. Some large diameter sausages are corded with many loops and 3 to 5 longitudinal strands, while others have only circular twine. Dry sausages are usually sold as new dry sausages (about 20 percent weight loss from original weight), moderately dry (about 30 percent of weight loss) and dry sausages (about 40 percent of weight loss).

Operations in Fermented Sausage Production

Choice of raw materials: The meat of adult, well-fed animals is preferred in raw sausage production. The use of chilled meat with a low pH is most suitable. For beef, the lowest pH of about 5.4-5.5 is obtained within a couple of days. Pork usually acidifies a little faster with final pH of 5.7-5.8. It is recommended to use meat that has been slightly frozen (-3° to -5°C) for one or two days prior to processing. The same holds true for fatty tissue. The use of beef is somewhat preferred for semidry sausages while pork seems to be more suitable for dry sausage manufacture. All meats used in raw sausages need extra careful trimming of fat and sinews; soft intermuscular fatty tissue in particular should be thoroughly removed.

Spices contribute mainly to the development of flavour in dry sausages but it has been shown that they also possess an inhibitory and stimulatory activity, thus influencing the growth of certain bacteria. Of the bacteria studied, Lactobacillus plantarum was found to be the species most frequently stimulated by different spices with respect to growth and to acid production.

Grinding, chopping and mixing: Methods of grinding and chopping vary widely with the different types of semidry or dry sausages. In a general way it can be said that the finer the degree of grinding and chopping the more complete will be the extraction of protein while the spreading or slicing properties of the finished product will be improved. The same is true for bacterial contamination. Beef is normally chopped first and then the pork components and other ingredients are added. Salt is added at the very end of chopping.

The chopped meat mass should be well mixed with other ingredients and should be either immediately stuffed or placed in shallow pans and held under refrigeration to enable the curing process and stabilisation of microflora. The meat mix in pans must not contain air pockets which are excluded by kneading; at the same time, the meat should be covered to eliminate its contact with air.

Stuffing: After remixing again in the mixer, the meat mix is packed in the stuffer as firmly as possible to exclude air pockets. Stuffing into casings should also be done firmly and carefully to exclude air. Air inside the casing will discolour the meat and reduce the shelf life of the sausage. After the sausage has been stuffed, the open end is tied (or clipped) and a loop is formed so the sausages can be suspended on rods. The air pockets under the casings are punctured wherever they are visible to eliminate air.

Raw sausages have historically been produced in natural casings. Many of them are identified by the casing used or by the manner in which string is tied around them. Casings, natural or artificial, should be prepared with great care. Sewn beef casings are also used for some large diameter semidry sausages.

Treatment prior to smoking. The stuffed sausages are placed in a special place or room to enable the surface moisture to escape. The optimum temperature of this preliminary drying varies from 20 to 23°C and the relative air humidity should not exceed 75-80 percent. The sausages must neither dry too fast nor retain surface moisture and become slimy. Another alternative is to hang the sausages overnight in a chiller at 6-8°C. If the sausage casings are overdried, smoke will not penetrate.

Smoking: Only hardwood sawdust smoke is applied. Uniform distribution of temperature and smoke throughout the smokehouse is essential. There is no standardised procedure for smoking different raw sausages. Sausages are smoked from several hours to several days or weeks according to the diameter of the product: small diameter sausages, from 10 to 40 hours; medium diameter sausages, 15-45 hours; and large diameter sausages, several days and exceptionally 1-3 weeks.

In traditional smokehouse, uniform distribution of smoke throughout the smokehouse can be achieved by the appropriate distribution of the fire pits in different parts of the smokehouse. In a modern well-designed smokehouse, the smoke density and its distribution are maintained at the required level with equipment provided for controlling circulation, temperature and humidity. While liquid smoke solutions are commonly used in many meat products, the use of liquid smoke sausage manufacture is not really consistent with traditional practices.

The smoking of semidry sausages differs from that of dry sausages. Semidry sausages are usually smoked at a higher temperature (22-32°C or above) and a heavier smoke is usually applied. During smoking or prior to smoking of semidry sausages the fermentation process is usually accompanied either by chance contaminants or by added starter cultures. By chance contamination, successful products are not always produced.

Quite often semidry sausages are subjected to temperatures of 50-60°C in order to improve flavour and to inhibit bacterial development. To avoid excessive shrinkage, smoked sausages are removed from the smokehouse as and when they are properly smoked. Different diameter sausages need different smoking times.

The smoking temperatures required for different dry sausages vary from 12-22°C but never exceed 30-31°C. A temperature of 14°C is considered as optimum in dry sausage smoking. Some producers in Europe or the USA do not smoke their dry sausages at all but keep them for at least ten days to several months in a room in which the desired air conditions are obtained by various combinations of temperature, relative humidity and air circulation. Some sausages are lightly smoked before drying.

Drying: Drying or ageing is a key operation, especially in dry sausage production. The drying rate for dry sausages should be as low as possible. The most critical point in drying is to avoid the pronounced surface coagulation of proteins and the formation of sausage surface skin. If the sausages lose moisture too rapidly during the initial stages of the drying period, the surface becomes hardened and a crust or ring develops immediately adjacent to the casing.

This hardened ring inhibits further transmission of moisture and the sausage has an excessively moist centre. Only a sufficiently wet and soft casing, a high relative humidity at the outset of the drying operation and the use of a lower relative humidity in the advanced stages of the process will permit moisture to migrate from the interior of the sausage into the outer layer. Thus, the sausage should dry from the inside outwards. If the outer layers of the sauage become hard, the diffusion of water will be inhibited and the sausage tends to spoil. In conclusion, if the drying rate is adequately slow the sausage casing will enable gradual drying. At the start of the drying or the drying and smoking process, relative humidity can be as high as 98 percent. In the following 2-4 days the relative humidity must gradually but slowly be reduced. In that manner, the drying rate at the surface will be kept as low as possible. It is advisable to regulate the relative humidity according to the decrease in water-activity value of the product. Thus, the air relative humidity should be approximately 4 percent lower than the value of water activity of the product. For instance, when the water activity value of the product is 0.96, the optimum relative humidity of the smokehouse should be about 92 percent, etc.

Too much humidity in the drying room favours the development of mould and sliming of products. Some producers do not object to a light white surface mould at the very beginning of the drying process. When fully developed, the mould is brushed off or, in some cases, washed. It is believed that mould contributes to the specific flavour of some products.

Sausages, produced in tropical climates, should be thoroughly dried, and smoked more heavily than products in the moderate climatic regions.

Sausage drying rooms should be equipped with a fan, and with facilities for dehumidification and chilling or warming the air, if necessary, as well as humidity and temperature control instruments.

DEFINITION OF RAW FERMENTED SAUSAGES

Raw-fermented sausages are uncooked meat products and consist of more or less coarse mixtures of lean meats and fatty tissues combined with salts, nitrite (curing agent), sugars and spices and other non-meat ingredients filled into casings.

They receive their characteristic properties (flavour, firm texture, red curing colour) through fermentation processes. Shorter or longer ripening phases combined with moisture reduction ("drying") are necessary to build-up the typical flavour and texture of the final product. The products are not subjected to any heat treatment during processing and are in most cases distributed and consumed raw.

Dried Meat Products

Definition of Dried Meat Products

Dried meat products are the result of the simple dehydration or drying of lean meat in natural conditions or in an artificially created environment.

Their processing is based on the experience that dehydrated meat, from which a substantial part of the natural tissue fluid was evaporated, will not easily spoil. Pieces of lean meat without adherent fat are cut to a specific uniform shape that permits the gradual and equal drying of whole batches of meat. Dried meat is not comparable to fresh meat in terms of shape and sensory and processing properties, but has significantly longer shelf-life.

Many of the nutritional properties of meat, in particular the protein content, remain unchanged through drying.

RAW MATERIALS FOR MEAT PROCESSING

The *two main components of processed meat products* are *animal muscle meat* and *animal fat*. Apart from pure muscle tissue, muscle meat also contains some connective tissue and inter- and intra-muscular fat, which determine the quality of muscle meat. Animal fats are of firmer or softer texture depending on their location in the animal body. In addition to the animal species, the texture of fats determines their processing quality.

Edible animal by-products such as skin, internal organs and blood also play a role as raw materials for meat processing. By-products are not generally used; they are part of specific processed meat products.

The first preparatory step for processing of meat into meat products is the product-oriented selection of raw animal materials, taking into account their quality and processing suitability and the characteristics of the meat products to be fabricated. Some meat products require lean meat without adhering fat or connective tissue, while others have a higher fat and/or connective tissue contents. Other products require firm animal fats, for others soft fats are better suited. Choosing appropriate raw materials is indispensable for efficient meat processing and is best done by visual selection and grading according to the tissue-specific properties.

Meat processors are advised to develop enterprise-specific standards of raw material composition for each meat product fabricated. The proper grading of raw materials, which needs skills and experience, has a decisive impact on the quality of the meat products and resulting revenues which can be generated.

For the needs of small to medium sized meat processing plants, simple grading schemes are described hereunder, with raw materials from pigs, cattle/buffaloes and other ruminants as well as poultry.

Selection and Grading of Manufacturing Meat from Pigs

The below proposal for selection and grading of pig meat refers to the utilization of the entire carcass for meat processing. Naturally, in many meat plants, valuable meat cuts may be excluded from further processing and marketed as fresh meat.

In these cases only the remaining carcass meat is used for further processing. Common cuts for fresh meat sales are tenderloin, loin, rump, the entire ham or parts of the ham (topside, silverside, knuckle) and parts of the neck and shoulder.

The proportion of carcass meat going into fresh meat sales or into further processing is decided by the operator on a case-by-case basis. If higher amounts of lean meat are required for further processing, more primal cut meat will be used for this purpose and vice versa.

Hereunder, a grading scheme for manufacturing-meat from pigs consisting of six grades is proposed. This standard can be refined or simplified as determined by consumer demand.

This meat is derived from body parts with large muscle groups such as loin (loin, tenderloin), hind leg (topside, silverside, rump) and shoulder. P1 meat is obtained during the preparation of choice cuts, when portions are trimmed off from the cuts for fresh meat sales.

Given the case that more of this P1 meat is needed for processing, some of the above mentioned cuts can be used completely for this purpose. Muscle groups with a high connective tissue content (neck, thin shoulder) are not considered suitable for P1. Grade *Pork 1* meat is used for whole muscle hams (raw-fermented, cured-cooked) and all products where the meat structure remains visible (coarse sausages, reconstituted hams).

Muscle Meats with Some Solid Fats

This manufacturing-meat category mainly comes from the leaner parts of the belly (near the loin) and pieces trimmed off from the hind leg. The fat content of grade P2 meat should not exceed 25%. The embedded fatty tissue must be firm and dry, as this meat is normally used for coarse products where meat and fat particles remain visible and can be sensed during chewing. For the same reason all visible hard and soft connective tissues should be removed. Typical products are fresh sausages, dry fermented sausages and luncheon meat and, to a certain extend, reconstituted hams.

Muscle Meat Trimmings

The meat trimmings for grade P3 can originate from all body parts, but its main source is the front quarter. As these meat trimmings usually contain smaller or larger amounts of soft connective tissue, they are mainly used as raw material for finely-chopped meat mixes. The hard connective tissue should be removed. The embedded fatty tissue can be of soft or firm texture, but its content must not exceed 10 % to allow for preparation of lean batter for further processing.

Pork Back Fat

The fatty tissue derived during cutting of pork carcasses and preparation of choice cuts can be divided into soft and firm tissue. The firm and dry fat for grade P4 is exclusively from the fatty layer under the skin on the backside of the pork carcass, hence the name "back fat". Back fat is primarily used as the fat portion of raw-fermented sausages and for the manufacture of finely chopped meat mixes of the raw-cooked type. For coarse fresh meat products, where usually P2 meat is taken, back fat in combination with meat P1 can be used instead, as with such a blend the same fat content as contained in P2 can be achieved.

Soft Fatty Trimmings

Apart from the firm fatty tissue (grade 4 pork), a variety of soft fatty tissues are obtained from the pork carcass. Because of their oily and wet appearance, they are unsuitable for manufacture of coarse products, but can be incorporated in finely chopped meat mixes of the raw-cooked type in quantities of up to 25% of the overall fat portion added. It can also be used as fatty material for some precooked-cooked products.

Pork Skin

Pork skin is normally used as food unlike other animal skins which are used in leather production. As the skin is exposed to contamination during slaughtering and cutting, special attention must be given to obtain pork skin of good hygienic quality. On its outer side pork skin should be free of hair and other impurities and on the inner side the connected fatty tissue must be thoroughly removed. Pork skin is collagen rich and in precooked form a valuable material for the manufacture of some meat products of the precooked-cooked type. Occasionally pork skin is also used in raw form in processed meat products normally shredded to small particles and either used as an ice/pork skin emulsion or as dry granules. However, this is an application for large industries. Pork skin can also be used for gelatine production.

SELECTION AND GRADING OF MANUFACTURING-MEAT FROM CATTLE

Similar to pork, valuable meat cuts (choice cuts) from beef are usually excluded from further processing and marketed as fresh meat. The most common fresh meat cuts are tenderloin, sirloin, topside, silverside, rump and parts of the neck and shoulder. The rest of the carcass meat as well as trimmings derived during the preparation of the above mentioned choice cuts are used as manufacturing-meat for all types of processed products.

The functional properties of beef are influenced to a large extend by the age of the animal. Meat from younger animals has a much higher water binding capacity than meat derived from a carcass of an older animal. For this reason meat from younger animals should be used for products requiring high binding and water holding capacity and meat from older animals is more suited for products undergoing a drying and fermentation process.

Similar to the grading scheme for pig meat, a simple scheme is proposed for the selection and grading of beef, which is considered suitable for small and medium operations. For beef, three grades of manufacturing meat are sufficient to cater for the needs of small to medium-size manufacturing. Beef fat and skin are usually not a raw material for meat processing.

The following grading scheme for beef is proposed:

Lean Muscle Meat

The meat is derived from the major muscles of the fore and hindquarter with the exception of shanks and belly muscles.

Meat parts used for this grade are mainly obtained as muscle trimmings from the manufacture of primal meat cuts and from smaller lean muscles which are not sold as special cuts.

For this grade, small meat trimmings removed from bones during deboning, flanks and shanks are used. As this meat is relatively high in connective tissue and fat, it is only used for the manufacture of finely chopped meat mixes. It is not suitable for use as coarse parts in meat mixes due to its tough texture.

In some regions, in particular in Asia, buffalo meat plays a major role in the manufacture of meat products often replacing beef. The proposed grading scheme for manufacturing meat from buffaloes is the same as for beef.

In the above illustrations only grade 1 buffalo meat is shown. Grade 2 and 3 for buffalo is similar as indicated for Beef 2 and 3. Buffalo meat has excellent properties for further processing, in particular a pronounced red colour, good water binding capacity and typical flavour. Differences in texture compared to beef (buffalo meat may be slightly tougher) play no role in further processing. Buffalo meat differs slightly from beef in terms of:

Colour: Buffalo meat has slightly darker red colour than beef, also processed meat products containing buffalo meat have a darker and more intensive red curing colour.

Taste: Buffalo meat has a more pronounced meat flavour and taste.

Texture: Buffalo meat cuts, upon ripening and aging, can be made sufficiently tender but remain with slightly stronger texture as compared to similar beef cuts.

Fat content: Buffalo meat is usually leaner than beef and the colour of buffalo fat is white as compared to the yellowish fat colour in beef.

SELECTION AND GRADING OF MANUFACTURING-MEAT FROM POULTRY

In global meat production poultry meat is taking the second place after pork. Due to its widespread availability and popularity and its mostly very competitive production cost, poultry meat has an increasing share as a raw material in processed meat. Turkey and chicken meat is very suitable for further processing purposes. Turkey meat, which has darker and brighter muscle components deriving from the same carcass is well suited for processed meat products. In some developed countries there are sizeable turkey meat industries, with outlets for processed turkey meat products, such as bologna/ frankfurters/ham sausage type sausages, and cooked turkey hams. Such products are similar to the equivalent ones fabricated with beef and pork, but they are usually leaner.

A widely practiced approach is to classify turkey carcasses in two grades. Grade A is top quality with no defects on the meat surface and general appearance. Entire frozen carcasses as retail goods belong to this category. Grade B is the lower category and this meat is usually taken for further processing. When producing turkey cuts, those cuts not needed or suitable for fresh meat sales, can also go into further processing.

In developing countries, the production of chicken meat is by far more important than the production of turkey meat. Chicken meat can be produced industrially around population centres and it is in high demand, particularly where pork is not consumed for socio-cultural or religious reasons.

The most popular processed products from chicken meat are *chicken frankfurters, hotdogs, chicken hams* and the various breaded and fried products of the *chicken nugget type*.

For the production of meat from chicken for processing, the same principles apply as in the beef and pork sector. Either the entire carcass meat is used for further processing, or some of the cuts go in fresh meat sales and the remaining into the manufacture. Chicken carcasses are usually cut in wings, legs and breast. Legs can be further subdivided into *thighs*and *drumsticks*. The breast consists of the larger *superficial breast muscle* and the smaller profound breast muscle, the latter is also called *"filet"*.

TRADITIONAL AND MODERN METHODS OF SAUSAGE FERMENTATION

In raw sausage manufacture, amongst the many factors involved, development of microflora and its effect on product quality plays a major part. The fermentation process in particular results in the desired flavour characteristics and tang.

In the traditional production process, fermentation is accomplished by natural flora. In order to achieve safe fermentation of the raw sausage, it is of importance to give the microflora the proper growth conditions as well as the appropriate type of the meat. One of a number of methods offered for choice are procedures requiring extended incubation times. For instance, the raw sausage mixture, containing meats, curing salts and sugar, can be placed in 15-18 cm deep pans and kept for 2-4 days at 3-4°C. After remixing, the mix is stuffed into casings and the drying process continued at 12-15°C with or without simultaneous smoking. A number of alternative procedures are found in practice.

The inherent bacterial sausage microorganisms use various sugar substances as energy sources, whereby they produce acids and contribute to the flavour of the raw sausages. While dextrose is degraded by almost all kinds of bacteria, lactose or starch products of higher molecularity are converted only by some of them. In the second case, the speed of acidification is delayed. Acidification, *i.e.*, with a sufficiently low pH is indispensable for adequate binding and colour development in the sausage. A pH value of 4.8 or below influences the taste, but does not contribute to better binding properties of the final product. However, this low pH gives fermented sausages excellent keeping qualities.

In general, addition of sugar varies from 0.3 to 2.0 percent. If dextrose or other easily degradable sugars are used, acidification is fast and the amount of sugar added should be somewhat lower. In opposition, corn syrup solid must be added at somewhat higher levels in order to compensate for its lower acidification properties.

A sufficient level of acidification can also be obtained by the addition of glutamine or some other acids and particularly by the addition of glucono-delta-lactone (GDL). When GDL is used, acidification (release of gluconic acid) occurs at 21-23°C. At GDL levels under 0.5 percent, the curing agent should be nitrate, but at GDL levels higher than 0.5 percent nitrite is preferred since too high a rate of acidification does not permit the process of denitrification performed by microorganisms.

Traditional fermentation of dry sausages in the pre-1940 period was performed as mentioned above by the inherent lactic acid-producing microorganisms or by back-inoculating a portion of recently fermented meat into the freshly prepared batch. However, this practice was associated with many doubts and uncertainties and to overcome them, the use of starter cultures was introduced in the early 1940s. The manufacture of fermented sausages changed essentially, particularly with the use of frozen concentrated starter cultures.

Bacterial starter cultures used in the production of dry sausages are lactic acid producing and belong mainly to the general Lactobacillus, Pediococcus and Streptococcus. Besides a few French (ferments lactiques) and American preparations of lactobacilli (Lactacell MC, etc.), and Spanish mixed-culture preparations, those mainly used on the European market are the micrococci and lactobacilli. As a matter of fact the market today is dominated by three types of products: starter cultures containing micrococci, starter cultures containing a mixture of micrococci and lactobacilli and starter cultures containing lactic acid-producing cocci. The reason for applying these bacteria in the production of raw sausages is their ability to produce a consistent and controlled acidification able to inhibit growth of undesirable microorganisms and as an aid in obtaining the desired structure and colour of the final product. Using starter cultures in current production practices, desired acidity can be achieved within 24 hours at a high (35-41°C) incubation temperature. This is in contrast to traditional manufacturing processes which utilised the natural flora of the meat as the source of lactic bacteria and required extended incubation times.

Storage. If stored in a warm hanging room, raw sausages shrink excessively and they become firm; if left in too humid or too cool a room, they soon lose their colour.

Most producers favour handling raw sausages, especially in warm weather, in temperatures around 18-22°C. This is done to avoid condensation which gathers on smoked sausages transferred from a cold to a warm place, such as the ordinary meat shop. Such condensation favours rapidly forming moulds.

Fermented Sausage Formulations

It is impossible to cover all the numerous types and potential variants of fermented sausages in this manual but it is possible to indicate some of the major products and suggest methods suitable for the manufacture of exclusive warm climate products. A point which is necessary to make clear is that every successful sausage manufacturer has to spend some time to adjust his sausage formulations to suit local requirements. Basic and supplementary spices used in semidry and dry sausage manufacture. The use of non-meat ingredients in fermented sausage manufacture is discussed under the heading "Non-meat ingredients".

Summer Sausages

Summer sausages are usually coarse-ground fermented products. They are produced under conditions that promote microorganism proliferation which imparts flavour and the desired keeping quality. Starter cultures may be helpful in obtaining quality products.

Formulation

- Basic ingredients for 100 kg
  - 55 kg lean beef
  - 25 kg pork bellies
  - 20 kg pork fat
- Characteristic seasoning formula per 1 kg
  - 30.0 g nitrite salt for curing
  - 0.5 g sodium nitrate
  - 3.0 g black pepper, coarsely ground
  - 1.2 g coriander
  - 2.0 g mustard seed
  - 0.4 gv garlic, fresh
  - 0.4 g allspice
  - 5.5 g dextrose
  - 2.5 g sucrose (some sausage makers prefer to have excess sugar present in final products which will produce the desired sweet flavour)

Casings

Pig casings: 35-45 mm in diameter

Processing and Handling

The meat and fat are ground separately through a grinder plate having 5 mm holes. After thorough mixing, the sausage mixture is stuffed into casings. The sausages are tied into the desired length.

Further handling of sausages may differ. Sometimes they are fermented at 38°C and 95 percent relative humidity for 1-3 days. Fermented sausages may also be manufactured using a commercial lactic acid starter culture or 0.6-0.7 percent glucono-delta-lactone. The drying process is conducted until the desired degree of quality is reached.

Air-dried Sausages

This type of sausage is distinguished by high quality characteristics. The particularly attractive appearance as well as aroma and keeping quality are attributable to a low pH and low water activity. These sausages are predominantly pork products and, if stuffed in wide casings, they are known as dry salamis.

The use of glucono-delta-lactone (0.3 to maximum 0.5 percent) can improve the texture and colour of the product and accelerate production but the quality of the finished product will be somewhat lower.

Formulations

- Basic ingredients for 100 kg
  - 20-35 kg beef, well trimmed, prefrozen
  - 30-45 kg pork (derived from older animals), well trimmed of fat, prefrozen
  - 20-30 kg backfat and pork belly
  - 20 kg beef, prefrozen at - 10°C
  - 35 kg pork, prefrozen at - 5°C, well trimmed of fat
  - 30 kg pork belly, prefrozen at - 5°C
  - 15 kg backfat, prefrozen at -5°C
- Characteristic seasoning formulae per 1 kg
  - 30.0 g nitrite salt for curing
  - 0.3 g potassium nitrate
  - 2.0 g dextrose
  - 4.0 g dry starch syrup
  - 2.5 g ground white pepper
  - 1.0 g white whole pepper
  - 0.5 g coriander
- 30.0 g nitrite salt for curing
  - 0.3 g potassium nitrate
  - 2.5 g dextrose
  - 2.5 g dry starch syrup
  - 5.0 g lactose
  - 3.0 g white pepper, ground

Casings

Corresponding diameters of artificial or natural casing

Processing and Handling

The beef is chopped in the cutter to approximately 2-3 mm grain size and at the end of chopping all seasoning ingredients except salt are added and homogenized. Pork components are placed at a low cutter knife speed and coarsely chopped. Kitchen salt must be added at the very end of chopping.

The stuffed and tied casings are kept at 20-24°C with a high relative humidity (92-94 percent) for 24 hours. During the next 24 hours the relative humidity is reduced to 90 percent. In the course of the following 3-4 days, the temperature is gradually decreased to approximately 14-16°C as is the relative humidity. The rest of the ageing process is carried out at 14-15°C and 76-78 percent relative humidity.

The uncontrolled growth of "wild mould" should be completely stopped for hygienic reasons. Should a uniform white surface coat of mould be desired, immersion inoculation with mould starter cultures must immediately follow stuffing.

Pepperoni

The name of this type of sausage is derived from the word "pepper", indicating that pepper is used in its spicing. Pepperoni are often produced as an air-dried product. They can also be produced either as all-pork or all-beef products.

Formulation

- Basic ingredients for 100 kg
  - 50 kg pork trimmings
  - 30 kg beef chucks, hearts, cheeks
  - 20 kg pork jowl.

Additional (non-meat) ingredients Addition of non-meat proteins, such as whey or soy products to pepperoni mixture is sometimes practised.

- Characteristic seasoning formulae per 1 kg
  - 28.0 g nitrite salt for curing
  - 0.5 g sodium or potassium nitrate
  - 0.5 g chili
  - 2.0 g allspice
  - 1.5 g fenugreek
  - 3.0 g ground pepper
  - 4.5 g red pepper
  - 1.5 g anise
  - 10.0 g sugar
  - 0.3 g peeled garlic
  - 2.5 g dextrose.
- 30.0 g nitrite salt for curing
  - 1.2 g monosodium glutamate
  - 4.0 g pepper
  - 2.0 g red pepper

  - 0.2 g chili
  - 3.5 g dextrose
  - 0.2 g garlic

Casings

Pig rounds: 28-32 and 35-44 mm in diameter

Processing and Handling

Pork and beef are passed first through a 10-12 mm plate and then run through a 3 mm plate. The mixing and seasoning of meat is performed in a mixer for at least 5 minutes. The meat mixture may be then either cured in pans, 15 cm deep, for two days at a temperature of 2°C or immediately stuffed. After curing, the cured mass is remixed and stuffed into casings without delay and linked in pieces of 24-25 cm. The linked sausages are dried or smoked and dried. Drying is carried out according to consumer taste.

If the mixture is not cured in pans but stuffed directly into casings, the sausages are held for about 10 days at 3°C prior to placing them in a green room at 22-24°C with a relative humidity of 80 percent for 48 hours. Smoking is performed at 32-34°C for 2-3 days. The sausages are further dried at 14°C.

An alternative method, especially if starter culture is added, is to ferment stuffed pepperoni mixture for 24 hours at 35°C and dry it at 12-14°C for 4-5 weeks or until desired firmness is obtained.

Pork and Beef Chorizos

Chorizos are a strongly hot spiced type of raw sausage, which can be sold and used as fresh (like merguez), semidry or dry sausages. Beef and fat, such as beef chucks, brisket fat and trimmed flank may be used in making an all-beef sausage variant.

Formulation

- Basic ingredients for 100 kg
  - 33 kg lean pork or lean beef
  - 33 kg pork neck or beef flank
  - 34 kg fat pork (jowl, belly, fat trimmings) or beef trimmings
- Characteristic seasoning formula per 1 kg
  - 28.0 g nitrite salt for curing
  - 0.4 g potassium nitrate
  - 0.8 g sugar
  - 1.0 g garlic, fresh
  - 2.0 g red pepper

- 1.5 g chili
- 0.5 g glucono-delta-lactone (optional)

Casings

Narrow and wide pig or sheep casings (28-38 mm)

Processing and Handling

The pork components are run through the 6-9 mm plate. If beef is used, it is ground more finely (3-4 mm plate). During mixing, curing salts and sugar are added and the mixture cured either in a shallow pan at 4°C or in casings after stuffing. Sausages are divided into lengths of 8-10 cm. After several smoking days (cold smoking at 14°C), the sausages are held at the same temperature until the desired texture is obtained.

An addition of 0.4 percent glucono-delta-lactone (if potassium nitrate has not been added in the mixture) will considerably accelerate the formation of desirable texture and contribute to a better keeping quality of the product.

Beef Salamis

There are several varieties of beef salamis, depending chiefly upon the length of the curing preblended sausage mixture, degree of fermentation, diameter of casings, etc.

The acceleration of beef salami manufacture may be achieved by adding glucono-delta-lactone or 0.5 percent of a starter culture in the mixture before stuffing.

Formulation

- Basic ingredients for 100 kg
  - 75 kg beef chucks or other beef
  - 25 kg beef brisket fat
- Characteristic seasoning formula per 1 kg
  - 25.0 g nitrite salt for curing
  - 0.4 g potassium nitrate
  - 0.5 g fresh garlic
  - 2.5 g ground white pepper
  - 2.0 g dextrose
  - 1.0 g red pepper
  - 0.2 g ginger
  - (lactic acid starter cultures may be added)

Casings

Beef middles or weasands.

Processing and Handling

The beef is run through a 4 mm plate of the grinder and the brisket fat may be diced. After mixing with other ingredients, the meat mass is placed in 15 cm deep pans for curing at 4°C for 15-48 hours. The mass is then remixed and stuffed into casings. Stuffing and tying must be very firm. The smoking is performed at gradually increasing temperatures, sometimes to 60°C or even more to stop the fermentation process. If a fully fermented product is preferred, the sausages should be hung for 3-5 days at 18-22°C, smoked in cold smoke for 2-4 days and dried according to demand. After smoking, the salami may be rinsed with water and then chilled.

To avoid the growth of mould, the sausages after removal from the smokehouse can be rinsed in hot water saturated with salt.

TURKISH AND ORIENTAL STYLE RAW BEEF SAUSAGES

In traditional Near East meat processing, several groups of sausages can be found but beef sausages or "sucuk" are doubtlessly more typical and frequent. Sucuks are popular Turkish national products but they are manufactured throughout the Near East by numerous local sucuk markers. This type of product substantially differs from its European counterparts. Sucuks are exclusively beef products but occasionally buffalo meat and, more rarely, mutton are used. Sucuks are sausages of roughly comminuted meat, mixed with fat and other ingredients, stuffed into sheep or beef casings and exposed to smoking and/or drying, including sometimes sun-drying. They are often annular shaped. There are at least several types of sucuks which can be classified according to proportions of fat and lean, diameter of casing used, degree of drying, etc. Sucuk production varies much depending upon the season.

Formulations

- Basic ingredients for 100 kg
  - 80 kg beef
  - 20 kg beef or mutton fat
  - 60 kg beef
  - 30 kg mutton
  - 10 kg fat
- Characteristic seasoning formula per 1 kg
  - 25.0 g salt or nitrite salt plus 0.5 g potassium nitrate
  - 2.0 g black pepper
  - 2.0 g red pepper
  - 1.5 g garlic, fresh

- 1.5 g cumin
- 0.3 g cinnamon
- 0.4 g ginger
- 0.2 g cloves
- 2.0 g sucrose

Casings

Animal casings, mostly beef and sheep, 24-26 mm or more in diameter

Processing and Handling

Prerigor meats are sometimes used in sucuk manufacturing. Meat grinding is very coarse (10 mm grinder plate is commonly used). All ingredients are thoroughly mixed together.

Casings for sucuk stuffing should be flushed inside with water just before use. Salted casings are washed free from salt before use and any casing that is left over should be resalted. Air-dried beef casings (air drying is the common method of beef casing preservation) must be free of any remnants of adhering fat.

Artificial casings are also occasionally utilised. The meat mass is sometimes chilled overnight and then firmly stuffed. The stuffed sucuks are often pressed under boards by means of weight.

The sausages can be smoked several days or several weeks. In fact, the drying process lasts 2-3 weeks or more, depending on casing size.

Air circulation is sometimes so high that sausages, particularly if the casings contain remnants of fat, very often acquire a slightly rancid tallow flavour. Intensive fermentation, which occurs rapidly during the first days of drying, nearly stops at the end of the process.

For more details on processing and sucuk handling, see the instructions for "Pork and beef chorizos". The addition of 0.4 percent of glucono-delta-lactone in the sucuk mixture can considerably reduce production time and improve the keeping quality of the finished product.

The usually accepted quality requirements for European semidry or dry sausages (flavour, colour) are often neglected by sucuk consumers. The most popular sucuk flavour is a specifically balanced mixture of fermentation products, added spices and components of fat and meat degradation. Very often the strongly accentuated garlic flavour and more or less fermented and, to some extent, rancid sucuk taste may be unpleasant to unaccustomed people.

Landjaegers

"Landjaegers" or hunter's sausages are the flat dry sausages, made by stuffing a mixture of beef, pork and seasonings into a medium diameter casing and cutting the stuffed casings in desired lengths. The characteristic operation is pressing, usually in wooden moulds, to flatten the sausage. A possible alternative in the production of these sausages for tropical regions is to use all-beef ingredients.

Formulations

Basic ingredients for 100 kg

a. 35 kg beef (may be held at - 10°C for 2-3 days)
   65 kg pork bellies (- 10°C for 2-3 days)

b. 70 kg beef
   30 kg pork backfat (may be held at - 10°C for 2-3 days)

c. all-beef landjaegers
   75-80 kg beef
   15-25 kg beef fat

Casings

Pig or sheep casings, 26-34 mm in diameter.

Processing and Handling

The beef is coarsely ground or directly chopped and then worked to a mass with usual additions (salt and seasoning). At the very end of chopping, the pork components are incorporated. The mix is stuffed into pig or sheep casings, cutting the stuffed casings off in about 30 cm lengths and then placing the sausages in wooden moulds.

The sausages are pressed into the moulds by hand where they take on their flat shape and each link is separated into two identical sausages connected together by a section of empty casing. Depending upon the force of pressure the sausages are produced in different thicknesses.

The landjaegers are allowed to remain in the press for 1-2 days at about 20-24°C or for about 2-3 days at approximately 18-20°C, giving sufficient time for biochemical and microbiological processes and for obtaining the required flat shape. The shaped sausages are then removed from the moulds, placed on smoke rods and hung for 3-4 days in a cold room (16-20°C) for surface drying and further "ripening". The sausages are then given a cold hardwood smoke overnight or a little longer.

From the smokehouse, the sausages are taken again to the drying room for half a day and then stored or packed.

Freshly stuffed sausages are 6 to 7 pairs to a kilo; and the finished product about 9 to 10 pairs to a kilo. Shelf life is at least 60 days.

EXAMPLES FOR CHICKEN CUTTING

Industrial Method

In large industrial operations, chicken are usually cut in the hanging position. Carcasses are suspended by the neck on a conveyor and pass though working stations. At each station a specific cut is made and a certain part removed, until finally only the bone-carcass structure remains.

The following is a widely practiced industrial cutting method: First the skin is incised around the body above the legs. Then the wings are loosened, by cutting between wings and carcass through the wing joint. Following minor incisions using a knife, the wings are pulled-off together with the breast meat. The legs are then pulled off the carcass and finally the fillets are removed. Only the bone-carcass structure with neck remains.

Small-scale Method

In small-scale operations more attention is usually paid to obtaining intact parts for individual sales. Conveyor systems are in most cases not available and therefore chicken carcasses are usually cut-up on a cutting board or table. Many different cutting styles have been developed. The following is one example.

First the chicken carcass is positioned on the cutting board with the breast muscle facing downwards. Then a deep cut is made just above the legs following the leg line. The two carcass parts are pulled apart and the legs are separated with a knife by splitting the backbone. Next the wings are cut off through the lower wing joint. If the breast and filets muscles are wanted separately, they are now removed from the upper carcass part and trimmed.

Lean meat can now be trimmed off the carcass parts for further processing; skin and fat are also obtained.

Grading of Chicken Meat for Large Operations

Chicken manufacturing-meat for larger processing operations is usually categorized in four different grades. The four grades of chicken meat are either used for pure chicken meat products or for mixed products. In Halal products made from red meat (beef, mutton), the fat portion may partially or fully be chicken skin.

Also in customary mixed red meat products (like frankfurter, bologna, breakfast sausages, luncheon meat, etc.) normally made of lean pork and beef or lean pork only, and pork fat, part of the lean pork may be substituted by lean chicken meat. This is usually done for cost-cutting reasons i.e. when cheaper chicken meat is available), but also to satisfy the growing demand for lower fat meat products.

The four grades are described below:

GRADE Chicken 1 (CH1) Chicken white muscle meat with visible fat, connective tissue and skin removed

For this grade mainly breast and filet meat is used. As meat of this grade is used for reconstituted chicken hams and chicken sausages with visible coarse meat parts, all fat and skin must be removed from the lean meat.

GRADE Chicken 2 (CH2) Chicken muscle meat with adhering subcutaneous and intermuscular fat

Deboned and skinless meat from all chicken cuts (breast, legs, wings) can be used. This meat is usually ground or chopped during further processing. Smaller quantities of subcutaneous and intermuscular fat are usually not removed and incorporated in the final product.

GRADE Chicken 3 (CH3) Chicken skin/fat

Chicken skin is removed from the carcass or individual cuts and collected separately. Chicken skin has a high fat content and is ground prior to being added to processed meat products. Chicken fat serves as the fat portion in all-chicken processed meat products such as chicken frankfurters or chicken bologna. It can also be used as fat in lean beef or mutton products, such as Halal frankfurters etc. Chicken skin is added to meat products for the same purpose as pork fat in pork/beef products, namely to contribute to product flavour and softer product texture.

GRADE Chicken 4 (CH4) Mechanically deboned chicken meat (MDM)

This grade is manufactured in industrial chicken plants by mechanically separating remaining muscle tissue from the chicken carcasses after removing legs and wings and the breast muscles including skin. Chicken necks are also used for MDM. MDM contains muscle meat, connective tissue and some fat remaining on the bones after removing the meat cuts. MDM is a typical industrial product and not produced in small operations. However, it is available on the meat market and can be purchased by smaller producers as frozen blocks for further processing.

Chicken MDM is an ingredient for lower-cost meat products for partial substitution of the lean meat. However, MDM addition is limited as high amounts of chicken-MDM will affect the quality of products (deficiencies in texture and taste) and may in some countries result in products which are not in line with national food regulations.

Grading of Chicken Meat for Small Operations

In small-scale operations, more emphasis is given to sales of fresh chicken parts. Therefore usually only three grades of processing meat are obtained:

GRADE 1: Trimmed lean breast and filet muscle meat (light colour)

GRADE 2: Leg meat (darker colour) and trimmings from carcass
GRADE 3: Skin/fat

CLASSICAL RAW-COOKED PRODUCTS

These are products of relatively high quality and good nutritional value as they contain meat and fat as the main components and water as the major non-meat ingredient. Ingredients of plant origin, if any, are used in small quantities and do not serve as an extender but as binders, e.g. soy isolate (approx. 2% added).

In contrast to the classical raw-cooked products, extended raw-cooked meat products are of inferior quality as they contain higher quantities of extenders and fillers of plant origin.

Principles of Manufacture

1. Extract and dissolve solid muscle proteins through physical (chopping) and chemical (salting) treatment of lean meat and blending with water.
2. Establish a network structure of liquid or gelatinous muscle proteins in the meat batter capable of surrounding and enclosing small particles of fat and binding water.
3. Stabilise the muscle protein network through heat coagulation by "cooking". Core temperature of products should be not lower than 70°-72°C.
4. Cool products down through immersion in cold water or cold water spray immediately after cooking, pass quickly temperature range from +40°C - +20°C, store in refrigerated rooms .

These are the essential steps in the processing technology for all types of raw-cooked meat products. If instead of classical raw-cooked products specific types of extended raw-cooked meat products are manufactured, the incorporation of the extenders and fillers chosen takes place at phase 2. Examples for extended raw-cooked products are Hotdogs or Luncheon meat.

For phase 1 and 2 of the manufacture of raw-cooked products it is essential that only raw (uncooked) lean meat and fat are used, as pre-cooked materials would not produce the necessary effects of protein solution, protein network building and increased binding of water. Protein solution and network building is achieved through a combination of effects:

- Mechanical cutting/chopping/comminuting of the muscle tissue resulting in the release and extraction of muscle proteins out of the muscle cells.
- Solution of part of the released previously solid muscle proteins in the water/salt mix added.

- Enhancement of the transition of solid muscle proteins to the gelatinous or liquid phase in the presence of phosphates or other suitable substances such as citrates.

At phase 3 the necessary heat treatment is applied. Through adequate heat treatment of the batter filled in sausage casings, cans, or portioned otherwise (e.g. meat loaves, meat balls), the heat coagulation of the liquid or gelatinous proteins in the viscous batter is effected forming a firm cross-linked network structure. This results in a rigid firm-elastic texture of the final product.

For the above reasons the term "raw-cooked" is used for this type of products as in the first phase "raw" processing and in the second phase "cooked" processing is taking place.

Raw Material and Additives

Lean meat is the principle raw material for classical raw-cooked meat products. It provides the muscle proteins, which play an important role in the processing technology of raw-cooked products. The lean meat component comprises mainly chilled *beef* and/or *pork*, but also *poultry meat* (turkey, chicken) is becoming increasingly popular for this purpose. Other meats like *mutton* and venison are less suitable and not commonly used. The same applies to muscular *slaughter by-products* such as heart, diaphragm and oesophagus.

Lean meat is used in quantities of 30-50% referred to the overall amount of batter for raw-cooked products. Beef forequarters are an important source of lean meat. As most of the lean meat will be finely chopped, also smaller meat pieces (trimmings) obtained during meat cutting can be used.

Visible fats and hard connective tissues should be removed from the lean parts. The lean meat must be thoroughly refrigerated. In order to quickly achieve a homogeneous batter, the chilled lean meat is pre-minced (3mm grinder disc) prior to comminuting in the bowl cutter.

Frozen lean meat should be "tempered", i.e. raising the temperature to make it softer, but keeping it below the freezing point (-1 - 5°C). It can then be placed frozen into the bowl chopper.

Beef, in particular from younger animals, has the best WBC. Good WBC is important to bind the amount of up to 30% of water, which is usually added to the batter to improve texture and palatability of the final products. Moreover, the lean meat itself has a water content of 70%, which also must be kept bound. The binding of all this water must remain stable in the sausage mix also during heat treatment, which is the hardest challenge for the water retention capacity. Low WBC results in loss of liquid during cooking visible as an accumulation of jelly under the sausage casing, inside cans etc.

Apart from the water, the meat batter must also retain the fat added. It is again the function of the proteins to keep the comminuted fat droplets dispersed in the batter also during heat treatment.

Fat, mainly chilled pork fat, but also chicken, beef or lamb fat or vegetable oil, is used in quantities of 15-30%. Not all animal fats are suitable for this purpose, as some would make the batter too "grainy" (kidney fat) and others too soft (intestinal fat). Back, belly and jowl fat (pork), skin (poultry) and to some extend intermuscular fat (beef) are suitable. Beef fat is generally inferior to pork and chicken fat and should only be used if pure beef products are to be fabricated.

Fat is an indispensable ingredient in classical raw-cooked meat products because they make products softer and more palatable and support desirable flavour. In order to achieve a good quality product with typical taste and texture, preferably fresh chilled fats should be used. Fats from long cold storage, in particular frozen fats, are less suitable, as they tend to become rancid. Fats should be free of hard connective tissue. The chilled fats are pre-minced (3mm grinder plate) for good immediate distribution in the batter. If vegetable oil is used instead of animal fats, this oil should be chilled but must remain liquid or viscous.

Other Animal Tissues

Animal by-products of the non-muscular type, such as internal organs, are not used for raw-cooked products. The only exception is animal skin.

Poultry skin, which has a high fat content, can be used as pork fat replacement or as the entire fat component, in particular in pure poultry products. Pork and cattle/veal skin do not contain fat, but connective tissue. Pork skin becomes gelatinous upon cooking and comminuting, and has the property to bind some water and produce a sticky elastic texture. It can be blended to the batter, and this mixture develops good cohesiveness upon heat treatment and cooling. Pork skin is an economic source of protein (nutritionally inferior to muscle protein) and may be used for lower quality and lower cost raw-cooked products.

In contrast to all other raw materials for this type of products, pork skin is often added pre-cooked, as pre-cooked pork skin is soft enough to be sufficiently comminuted by the equipment used. Pre-cooking also reduces the high bacterial content of pork skins. In some cases raw-frozen pork skin is finely chopped with ice flakes and added. Poultry skin, used for specific products as a carrier of larger quantities of fat, is used uncooked.

Water is added in quantities of approximately 15-35%. It makes products juicy and easily chewable. Water is not primarily used as "filler" in raw/cooked meat products. It is absolutely necessary as a carrier and solvent for the muscle proteins. Water together with salt and phosphates, which enforce the process, is indispensable for the desired extraction and water binding of the muscle proteins.

As the protein extraction is best under low temperatures, water is often added frozen as ice to keep the temperature of the meat batter low. Low temperatures are also needed to avoid excessive rise of temperature in the area close to the fast rotating bowl chopper knives. Excessive temperatures there would spoil the ability of affected proteins in water binding and gelation. For quick and even distribution (in order to achieve an instant cooling effect in all parts of the chopping bowl), small-size ice flakes should be used or ice blocks should be crushed into small pieces before being added to the batter.

Additives and spices (salt, phosphates, ascorbic acid, various ground spices) are used in dry-powder form and must be stored in a cool and dry place in closed containers and frequently checked for impurities. These additives are used in smaller quantities and need to be accurately weighed to avoid under- or overdosage.

Salt and/or nitrite curing salt serve for salty flavour, solution of muscle proteins and curing colour. Approximately 2% is normally applied.

Phosphates assist in solubilizing muscle proteins. Approximately 0.3% is normally used.

Ascorbic acid acts as a catalyst in the curing process. Approximately 0.05 – 0.1% is normally used.

Note: In the process of chopping, the curing substances get homogeneously blended to the batter so that equal curing is secured. In order to accelerate this curing process, ascorbic acid (or sodium ascorbate or sodium erythorbate) is added to boost the reaction of myoglobin with curing salt. Ascorbic acid reacts instantly upon contact with the nitrite. In order to avoid a premature reaction before the nitrite is homogeneously distributed in the batter (which would result in unequal curing), it is recommended to add the ascorbic acid in the final phase of the comminuting process.

Spices may vary, but pepper (in combination with salt) forms the basis for a raw-cooked product seasoning mixture. Other spices are supplementary and are chosen according to the type of the product and local preferences.

Additional measures for proper cohesiveness and water binding

The protein network in the final batter will be the more complete and functional as more meat protein is used in the mixture. To further facilitate the formation of the protein network, where the meat content is low, the following measures can be taken:

- Addition of small amounts of proteins (approx. 2% of dry substance) mainly originating from soy beans (isolated soy protein) or derived from milk (milk protein, caseinate) or blood (plasma protein) can be added. They increase the amount of available proteins resulting in more comprehensive protein network structures.

- Lean beef taken from the carcass up to 6 hours after slaughter or pork up to one hour after slaughter ("pre-rigor meat") is still high in pH (approx. 6,5) and contains natural phosphate (ATP) which develops the same effect as synthetic phosphates which are commonly used. This method is not frequently practised as in most cases pre-rigor beef is not readily available. Where pre-rigor meat can be made available (e.g. where slaughter and further processing operations are carried out in the same meat plant), this method should be considered.

Technological Procedures

Meat processing techniques changed over the years following improved equipment designs and performances. The methods described below take into account the performance of the equipment expected to be accessible to small or medium scale meat processors in developing countries. The following processing steps achieve satisfactory final products:

The Chopping Process

Step 1 : The lean meat portion is pre-minced and kept chilled (0ËšC).

Step 2 : The lean meat is placed in the bowl cutter with the salt and additives for the whole batch. The mixture is chopped for 5-10 rounds without ice. This step is called "dry-chopping".

Step 3 : Ice is added and the chopping continued at fast bowl chopper speed until the ice is evenly incorporated and a "sticky" lean batter is achieved. Now also the spices are added.

Step 4 : The fat (pre-minced and chilled) is added and the mixture chopped at high speed until a homogeneous batter made of lean and fatty tissues is achieved.

Step 5 : The final temperature of the batter should not exceed +12°C

The extraction of muscle proteins can be further enhanced and a uniform distribution of all components in the mixture accelerated by:

- Sharp bowl chopper knives, high rotation speed (up to 7000 rpm)
- Specific knife shapes
- Increased number of knives (up to 8)
- Specific knife positioning for improved cutting sequence
- Passing the batter through emulsifying machines.

Not all raw-cooked products are entirely made of finely chopped batter. Many products are made of a mixture of finely chopped batter and coarse meat pieces. The coarse meat material can be cut by hand (strips of meat with

1-2 cm diameter) or pre-minced (5-8 mm hole size in grinder disc). It is usually pre-salted and spiced before incorporated in the finely chopped batter. Bigger meat pieces are separately cured (preferably in curing brine over 24 hours) and should be manually massaged or tumbled to extract protein from the surface of the meat pieces, which after heat treatment, makes the coarse pieces adhere firmly to the surrounding batter

Filling the Product Mixture into Casings or other Containers

The product mixture is usually filled into casings or other containers. Casings can be natural or made of different materials and are available in different shapes and calibres. Most casings require pre-treatment before being used.

The product mixture can be manually filled into casings using a simple funnel, but even in small-scale meat processing operations simple filling machines are commonly used. Before and during filling the following aspects are of specific importance:

- The filling machine must be clean and properly assembled (check piston gasket, attach funnel suitable for the calibre of the casing = rule of thumb: diameter of funnel to be $^2/_3$ of casing)
- The mixture must be placed tightly in the cylinder. Large enclosures of air must be avoided as they will burst during filling and produce loosely filled casings or lead to air pockets in final products.
- Casings should be filled to their maximum capacity to avoid surface wrinkles in the final product.

Sausages of smaller calibres are filled into longer strands of natural (mostly sheep) or collagen casings. These strands are divided into portions by manually or mechanically twisting or tying with a string. Sausages of bigger calibres are stuffed into individually cut casings. Both ends are tied with a string or sealed with a metal clip .

Using the filling machine, portions of the product mixture can also be filled in cans and glass jars. Meat loaf moulds are usually filled by hand.

The period between producing the batter and filling it into casings, cans, moulds etc. should be kept as short as possible. In the batter, even under relatively low temperatures, acid producing microorganisms can develop. The result is the lowering of the pH and reduction of the water binding capacity (WBC) in the batter.

METHODS OF HEAT TREATMENT

In order to achieve:

- firm texture

- intensive curing colour
- microbiological stability for cold storage and
- palatability for consumption

Raw-cooked products undergo specific heat treatments as the final processing step, which vary according to the type of products. The most complex heat treatment is applied to sausages. For sausages, after filling in casings, a combination of heat treatments, usually carried out in three steps (*reddening, smoking, cooking*) is applied. In most small and medium scale operations simple individual machinery (smoke house, cooking vat) suffices. In larger operations more sophisticated equipment (computerized smoking/ cooking chambers) is used.

The following describes a typical procedure for sausages undergoing all three steps using simple equipment:

Reddening: After filling, the sausages are hung on sticks in a way that they do not touch each other, thus allowing for air circulation around the individual pieces. The sticks are transferred to the pre-heated smokehouse and exposed to modest heat treatment (hot air) in a first phase without smoke at approximately +50°C. At this temperature the curing process in the product mixture is accelerated and completed within a relatively short time depending on the calibre of the sausages (from 15 minutes for calibre 22 to one hour for calibre 90). The presence of cure accelerators (ascorbic acid, sodium ascorbate) in the batter enhances this curing reaction. Hot-smoking: After the sausages to be smoked have gone through reddening and developed a red curing colour, the process of hot smoking is initiated. In small and medium scale operations sawdust and open gas flames or electrical heating are used to generate the smoke. Besides these simple smoking facilities, more advanced methods are also available.

The sawdust should originate from untreated wood (preferably hard wood) and must be free of any impurities. Care must be taken that the smoke is developed through smouldering. Open flames must be avoided. Raw-cooked sausages are hot-smoked at temperatures of +65 to 70°C until a desired product colour is achieved (30-60 minutes). As the smoking is carried out at relatively high temperatures, it can be considered part of the overall heat treatment of the product.

Note: The main purpose of sausage smoking is to provide a smoke flavour colour. The aspect of product preservation is only secondary. Smoking with gaseous smoke is only applicable, when sausages are filled in smoke-permeable casings (natural, cellulose or collagen casings). In some advanced smoking/cooking chambers "smoking" can also be carried out by spraying liquid smoke particles onto the product during cooking.

Alternatively, raw-cooked sausage products in smoke impermeable casings can also be produced with smoke flavour. In this case liquid smoke flavour is added to the batter during chopping to achieve the desired flavour.

Cooking: After the hot-smoking is completed or for un-smoked products after the reddening phase was initiated, the sausages are taken for further heat treatment ("cooking", "scalding") in order to achieve complete protein coagulation.

In small and medium scale operations the sausages are transferred to a cooking vat and submerged in hot water of +74° to +80°C for a certain period of time until a core temperature of at least +72°C is achieved. As a rule of thumb for products exposed to moist heat, 1 minute of heat treatment at +70°C per 1 mm of sausage diameter is required (e.g. cal. 60 needs 60 min.).

During the cooking process all products must be covered by hot water to avoid discolouration and partial undercooking. A floating grill can be used. When a smoking/cooking chamber is used, the sausages remain in the chamber after smoking and cooking is done by applying steam or water vapour saturated hot air. Cooking is essential and indispensable for all raw-cooked meat products to build up a strong complete network of coagulated protein and make the products "elastic". It also reduces the bacterial load present in the raw batter. The following facts are important to note:

- Core temperatures of +65°C would be sufficient to achieve the required texture through protein coagulation. However, for hygienic reasons and in order to eliminate a major part of the micro-organisms present in the batter, core temperatures of +70° to +72°C must be reached.
- Core temperatures up to +78°C are recommended for production in tropical countries in view of the ambient hot environment during handling and transport.
- Core temperatures above +78°C should be avoided as this would negatively affect the texture of the products (atypical soft).

Cooling of Cooked Products

After sufficient heat treatment was applied to the sausages to achieve the desired texture, colour and flavour, care must be taken to quickly reduce the product temperature through cooling. Products should never be kept or stored in the temperature range of +20°C to +40° as this would stimulate the growth of remaining bacteria/spores. The rapid cooling is practically achieved by immersing the products in a container of cold water or in modern cooking chambers by showering with cold water. After dropping the product temperature and superficial drying of the sausage surfaces, the products must be immediately transferred into chillers where they can be stored for a few days to a few weeks depending on the type of product.

Mode of Consumption

Raw-cooked sausages of smaller diameters such as frankfurters, hotdogs or Viennas are mainly consumed hot. They are heated up immediately before consumption. These small calibre sausages are mostly filled in edible casings and the edible casings are consumed as part of the product.

Larger calibre sausages are mainly eaten cold on sandwiches. The casings are removed and the sausage is cut into thin slices.

Raw-cooked Products others than Sausages

Apart from using casings the batter can also be filled into containers such as tin cans or glass jars for subsequent sterilization, resulting in canned products of the raw-cooked type. The use of special moulds is also common for baking in ovens or in hot air. Such finished products are called meat loaves. The batter can also be shaped as meat balls, which are directly submerged into hot water. Meat balls are popular as ingredients for soups.

Theory of Formation of "Raw-cooked" Meat Mixes

The main components of meat mixes for raw-cooked meat products are animal protein, animal fat and water. For raw-cooked meat mixes it is essential that fat and water components are evenly dispersed as tiny droplets and are kept stabilized in a finely comminuted protein mass ("protein matrix"). Keeping the fat and water droplets "stabilized" in the protein matrix means to prevent their confluescence to larger drops. In order to achieve this target, specific conditions must be met with regard to raw material selection, comminuting equipment and techniques and mix temperatures not to exceed during comminuting.

Through the comminuting process, the tiny lengthy structures of the muscle fibers are cut into a multitude of small fragments. As a result the three types of proteins present in the muscle tissue are set free. These are

- connective tissue proteins, mostly collagen, deriving from cell membranes and intercellular tissue,
- water-soluble soft sarcoplasmatic proteins from inside the muscle cells,
- myofibrillar proteins, which are solid protein chains (responsible for muscle contraction) inside the muscle cells composed of actin and myosin proteins.

Myofibrillar (*actin* and *myosin*) proteins play a decisive role in formation of raw-cooked meat mixes. Extracted from the muscle cells by comminuting the muscle tissue, these proteins are capable of *absorbing water* and *swelling in volume*. Myofibrillar proteins are *"salt-soluble"*, which means that in the presence of salt (NaCl) and water, they can be transferred from the solid to a gelatinous or liquid phase. This does not mean that all myofibrillar proteins present in the mixture are subject to gelation or are fully solubilized. The degree of gelation or solubilization depends on the amount of salt available, intensity of comminuting, pH of the meat and the processing temperature. Hence the mix contains different structural phases of myofibrillar proteins:

- one fraction are muscle cell fragments with unchanged solid myofibrillar proteins,
- other myofibrillar proteins are swollen through uptake of water and
- a substantial part of the myofibrillar proteins become gelatinous once fully solubilized through the impact of salt and water.

7

Meat Packaging Technology

MEAT PACKAGING

Packaging is a scientific method of preserving the quality and safety of the food it contains from the time of manufacture to the time it is used by the consumer and also to display the product in the most attractive manner for consumer preference.

The ability to convince a potential buyer to use a given product has always been influenced by product packaging and has become an important marketing tool for communicating brand values. Besides protecting the product from physical, chemical, or biological damage, it should also enhance marketability of the product and be environment friendly. The only criterion customers have at the point of purchase of fresh meat cuts is visual appearance. Consequently, desirable meat colour must be maintained during chilled storage, distribution and subsequent retail display if preservative packaging systems are to be effective. In the processed meat products, other factors such as dehydration, lipid oxidation, discolouration and loss of aroma must be taken into account.

Packaging of meat and meat products has always been challenging because of perishable nature of the product. Today, packaging constitutes an important department of a food processing industry and many meat-packaging systems are available with different attributes and applications. These systems range from overwrap packaging for short-term chilled storage and retail display to 100 per cent carbon dioxide atmosphere packaging for long-term chilled storage. Modern meat packaging technologies include vacuum packaging, modified atmosphere packaging, controlled atmosphere packaging, active packaging, smart packaging, etc. These strive to enhance the food safety and quality in as natural way as far as possible. The trends in innovative packaging techniques are discussed in the text that follows.

Packaging Forms and Materials

Packaging constitutes both flexible and rigid materials, alone or in combination with other preservation methods, to offer the necessary barrier, inactivation, and containment properties required for successful food packaging. Examples of typical flexible packaging materials include polyester, polypropylene, polyethylene, polyvinyl, polyolefins, vinylidene, vinyl chloride, surlyn, or nylon, which provide the needed characteristics such as strength, permeability, and saleability in the food industry. Additionally, flexible packaging materials can be combined with gaseous environments to extend the shelf life of many foods, including meat and poultry. Besides synthetically derived packaging materials, flexible packaging can encompass the use of edible films, gels, or coatings made from polysaccharides, proteins, lipids, or composites of any or all three. The benefits of using edible films as packaging materials are threefold: these films may resist the migration of outer moisture into the food during storage; films may serve as gas and solute barriers; and films compliment other types of packaging by improving the quality and shelf life of foods. And finally, the combination of rigid packaging materials made from metal, glass, or plastic and combined with heat, provides the most effective and widely used method of food preservation. The various traits of materials used in packaging of meat. Nearly all-available films for meat packaging are of thermoplastic materials, and therefore heat seal able, resulting in hermetically sealed plastic pouches, bags, etc. A high oxygen barrier is important in the application of films for packaging meat and meat products. Films made of polyvinylchloride (PVC), polyethylene (PE) or polypropylene (PP) have a relatively high oxygen permeability, whereas polyvinylidene chloride (PVDC), polyester (PETP), polyamide (PA) and cellulose film (ZG) are less or almost non-permeable to oxygen. The latter materials are therefore better suited for packaging meat and meat products. However, the materials of the first group are frequently used as laminates with materials of the second group in order to achieve special effects regarding mechanical strength, heat sealing properties or making the package practically impermeable to both oxygen and water vapour.

PACKAGING TECHNIQUES

Choice of packaging systems and the use depend on many factors including cost, the volume of product being handled, and presentation. Some of the recent methods used in meat industry are:

Vacuum Packaging

Vacuum packaging refers to packaging in containers (rigid or flexible), from which substantially all air has been removed prior to final sealing of container. This method of packaging is actually a form of "modified atmosphere" since normal air is removed from the package. The various advantages of vacuum packaging are that these extend shelf life by controlling oxidative rancidity and by preventing the growth of normal spoilage bacteria (aerobic organisms such as pseudomonas are suppressed and lactic acid bacteria are favoured which can grow to high numbers without causing spoilage), reduce freezer burn and loss of moisture at product surface, require minimum storage space and Leakers are easily detected (small puncture or pin hole in a vacuum pack is easily detected by looking for loose packages).

Today vacuum packaging is the most widely used method employed to extend the storage life of fresh meat. The air within the package must be evacuated to normal anoxic levels, *i.e.*, less than 500ppm to prevent browning due to residual oxygen. The exclusion of oxygen from meat surface immediately after carcass is broken into primals preserves the meat's potential to re-oxygenate soon after retail pack display. The preservative effect of vacuum packaging is by inhibition of potent spoilage bacteria at normal meat pH under optimum vacuum packaged conditions. The microenvironment within the pack will determine the type of micro-flora, which develops. When meat is first vacuum packaged any residual oxygen remaining in the pack is consumed by meat and muscle pigments and CO_2 is produced as an end product of tissue and microbial respiration. Under a good vacuum the package head space consists of <1 per cent O_2 and 10-20 per cent CO_2 which restricts the growth of aerobic microorganisms such as pseudomonas and favours desirable facultative anaerobic organisms such as lactobacillus but also undesirable Bronchothrix thermosphacta. These slow growing bacteria cause less offensive types of spoilage even at higher bacterial numbers than other aerobic organisms and will eventually cause spoilage but only after many weeks of storage. If vacuum packaging procedures are followed correctly, the storage life of meat can be extended.

Care must be taken to keep initial bacterial load as low as possible by following strict hygienic procedures. The temperature must be maintained at or near 0-degree Celsius. High pH and DFD meat must be avoided as early spoilage may occur. High vacuum levels result in the most desirable muscle colour and fat appearance. Vacuum packaged meat of normal pH (<5.8) can be stored for 12 to 14 weeks at 0°C. However, vacuum packaging is considered unsuitable for red meats for retail display purposes since the oxygen-depleted atmosphere causes the meat in these packages to appear purplish in colour and therefore not acceptable to consumers. Vacuum packaging is ineffective for whole carcasses or cuts which prevent the packaging film being closely applied to all surfaces. There are four basic ways; the first one involves the use of heat shrinking flexible packaging material around the primal cuts. When exposed to heat it shrinks and increases film thickness, improves mechanical resistance and reduces the drip loss. The second type involves the use of preformed plastic bags/pouches made up of polyamide (PA) as the outer layer, while the inner core is polyethylene (PE). Polyamide provides barrier properties and physical strength while polyethylene provides sealing properties. The third type provides the use of thermoforming trays in line from a base web. Here the product is placed into the tray and then a film web is allowed to cover the tray from a second reel of film followed by pack evaluation and vacuum sealing. The fourth type is vacuum skin packaging, in which the meat is placed in a rigid preformed tray or on the flat surface of a flexible base material followed by heat softening, evacuation and vacuum shrink sealing.

Gas Packaging

Gas packaging can be defined as the alteration of the proportional volumes of the gases, which comprise a normal atmosphere. This type of packaging generally falls into two categories:

1. *Modified Atmosphere Packaging (MAP):* Refers to enclosing a product in some type of barrier and modifying the atmosphere either by drawing a vacuum or filling with a gas mix. During storage the level of these gases will change due to respiration of the product and permeability of the film.
2. *Controlled Atmosphere Packaging:* A system where gases are added or removed to maintain a desired balance is called controlled atmosphere packaging. In this case the bulk bin or the storage vessel is virtually impermeable.

ADVANTAGES OF GAS PACKAGING

Advantages of gas packaging are that it allows meat to bloom (if the oxygenated red colour is desired, blend incorporated oxygen can be used to control colour), extends shelf life by controlling oxidation by displacing the oxygen with CO_2 and N_2 inhibits growth of normal aerobic spoilage organisms besides surface molds, prevents crushing of soft products as gas mixtures of CO_2 and N_2 equalise the pressure in the package and prevent package collapse and retains moisture but water can move from centre of product to surface resulting in staling.

In MAP systems each gas has a specific role to play in extending the shelf life and maintaining the appearance of packaged meat. The most commonly used gases for the packaging of meat is CO_2, N_2 and O_2 although other gases including CO, N_2O, Argon, SO_2 and Ozone have been tried to a limited extent.

The most commonly used gas mixture for fresh red meat is high O_2, which has a minimum 60-70 per cent O_2 and 30-40 per cent CO_2. Packaging in high oxygen extends the time for occurrence of microbiological spoilage and discolouration of meat. Most chilled poultry products are sold in prewrapped O_2 permeable film, which prevents moisture loss and the spread of contaminating microorganisms. However, where O_2 impermeable films are used, mainly for duck and turkey, the benefits are in extended shelf life.

There are also reports that high CO_2 atmosphere packaging of poultry extends the storage life up to three times that of storage in air. The modified atmosphere packaging for cured meat products has generated certain problems viz colour stability of cured meat products stored under illumination during retail display. Light exposure including intensity of light or illuminance, oxygen transmission rate of packaging material and residual level of O_2 have been found to effect the colour stability of cured meats.

Discolouration in controlled atmosphere packaging can be prevented only by exclusion of all O_2 from the package, which requires special evacuation equipment and totally gas impermeable packaging materials. If CO_2 is a major or sole component of the input atmosphere, then the quantity of added gas must be adjusted to assure that the intended atmosphere persists after dissolution of the gas into the product. For chilled meat, the most effective technology to date is the high CO_2 controlled atmosphere packaging system. This regime limits microbial deterioration through optimal storage temperatures (-1.5°C), high levels of CO_2, low residual O_2 (=0.05 per cent), and use of a gas impermeable films.

Active and Intelligent Packaging

An active or intelligent packaging is a packaging, which has an additional desired role that improves the quality of the food product as a supplement to pure packaging properties. These properties can either be integrated in the packaging material or separated from it. Intelligent packaging contains an external or internal indicator for the active product history and quality. It has an inherent ability to gather information on its operating environment or history, to process that information in order to draw intelligent inferences from it and to act on those inferences by changing its characteristics in an advantageous manner. Active packaging has an extra function in addition to that of providing a protective barrier against external influence. It can control, and even react to, phenomena taking place inside the package. It actively and constantly changes either package permeation properties or the concentration of various volatiles and gases in the package headspace during storage, or actively adds microbial, antioxidative or other quality improving agents, *e.g.*, flavour enhancing substances, via packaging material into food in small amounts during storage. Examples of active packaging are those that exhibit preferential permeability; temperature compensating attributes; oxygen, carbon dioxide, alcohol, or ethylene scavenging systems; moisture absorbers; antimicrobials incorporated directly into the packaging matrix; controlled release of colourants, flavours, spices, minerals, antimicrobials, ethanol, or antioxidants; or the ability to absorb odors. Based on their way of functioning, active packaging are categorised into four groups viz scavengers, emitters, indicators and others.

The first three types can be evaluated as groups, because of the similarities in the intended function of the product but for other types (*e.g.* Susceptor materials) of active packaging, an individual evaluation will have to be done from case to case.

Active Packaging Scavengers

Active packaging scavengers consist of ingredients, which are intended to absorb, remove and then eliminate substances, such as oxygen, ethylene, moisture, or taint from the interior of a food package. The constituents of active packaging material and articles have an effect on the shelf life or the organoleptic properties of the food. The examples are the oxygen scavengers and ethylene scavengers, which are used for eliminating oxygen and ethylene respectively from the package resulting in an extended shelf life of the food. Commercial oxygen scavengers are in most cases based on iron powder, which, in the presence of oxygen, forms iron oxide. Non-metallic and organo-metallic compounds have also been developed.

Active Packaging Indicators

Various indicators, *viz* indicators for temperature, microbial spoilage, package integrity, physical shock, and product authenticity etc can give information on the quality of the food product directly, besides the package and its headspace gases, as well as on the storage conditions of the package. Some indicators do not need to interact with the product or the headspace, while others do. These indicators are often called intelligent packaging and certain concepts are already commercially available, and their uses seem to be increasing. New concepts of leak indicators and freshness indicators are patented, and it can be expected that new commercially available products will be available in the near future.

Antimicrobial Agents

Quite recently, the idea of incorporating antimicrobial agents directly into packaging films has been developed. Thus, the packaging material can serve as a source of releasing preservatives or antimicrobial agents, or even prevent the growth of microorganisms. Flores and others (1997) reviewed the products and patents in the area of active packaging and identified antimicrobial packaging as one of the most promising versions of an Active Packaging system. Of the active packaging applications, the incorporation and/or slow release of antimicrobials is receiving considerable attention as a means of extending the bacterial lag phase, slowing the growth rate of microorganisms, prolonging shelf life, and maintaining food safety. The general premise of antimicrobial packaging is a controlled migration of the compound to the food through diffusion or partitioning (Han, 2000), which not only allows for initial inhibition of undesirable microorganisms, but also retains residual activity over time.

Active Packaging Emitters

This group of active packaging contains, or produces, substances, which are meant to migrate into the food packaging headspace or into the food in order to obtain a technological effect in the atmosphere in the packaging or in the food itself as, *e.g.*, food additives, flavourings or biocides. In these cases, the consumer together with the food ingests the components.

Edible Films

Considerable research has been conducted to develop and apply bio-based polymers mainly from a variety of agricultural commodities and/or waste of food products to reduce the amount of synthetic polymer waste. Edible coatings and films offer a variety of advantages such as biodegradability, edibility, biocompatibility, aesthetic appearance, and barrier properties. The benefits of using edible films as packaging materials are threefold. Since the water activity (aw) is critical for microbial, chemical, and enzymatic activities, edible films may resist the migration of outer moisture into the food during storage.

They may serve as gas and solute barriers and compliment other types of packaging by improving the quality and shelf life of foods. With the incorporation of antimicrobials into the matrix, edible gels, films, or coatings can exhibit active packaging characteristics, which are attributes necessary for controlling undesirable microbes in muscle foods. Antimicrobial compounds such as organic acids (acetic, propionic, benzoic, sorbic, lactic, lauric), potassium sorbate, bacteriocins (nisin, lacticin), grape seed extracts, spice extracts (thymol, p-cymene, cinnamaldehyde), thiosulfinates (allicin), enzymes (peroxidase, lysozyme), proteins (conalbumin), isothiocyanates (allylisothicyanate), antibiotics (imazalil), fungicides (benomyl), chelating agents (EDTA), metals (silver), or parabens (heptylparaben) can be added to edible films to reduce bacteria in solution, on culture media, or on a variety of muscle foods. The incorporated antimicrobial agents must be edible food grade compounds since they have to be consumed along with the edible gels, films, or coatings.

Biopolymers

Research efforts since 1940s have been devoted to develop bio-based polymer films derived from natural sources such as plant and animal materials. The use of plant material to form films is an active research topic. The advantages of using biopolymers for food packaging include: reduced dependence on petroleum-based packaging, use of a renewable agricultural resource, the biopolymers can act as carriers to deliver shelf-life extenders such as antimicrobials or antioxidants, and biodegradability.

Since the forming of a film requires cross-linking of molecular units to impart strength and flexibility, proteins and carbohydrates are often the best candidates for biopolymer films. Proteins investigated include wheat gluten, corn zein, whey, pea protein, meat proteins, egg proteins, and soy. Starches studied in the literature include alginate, polysaccharides, cellulose, carrageenans, and microbial polysaccharides. Pearce and Lavers reported using carrageenan to protect frozen poultry and Klose *et al.* incorporated an antioxidant into a gelatin coating to slow the development of oxidative rancidity in cut poultry meat prior to freezing. Edible film packaging of meat act as a moisture barrier, oxygen barrier, texture modifier, breading adhesion aid, mold suppressor, bacterial inhibitor, physical protectant, oil barrier, antimicrobial carrier and antioxidant carrier. The colour stability of freshly cut beef surfaces has been reported to improve by exposing the meat surface to antioxidant-impregnated corn zein films.

MEAT-PACKING INDUSTRY

Meat-Packing Industry, large industry involving the slaughtering, processing, and distribution of cattle, sheep, and hogs. It is one of the most important industries in the U.S., and has its primary concentration in the Middle West. The packing industry has tended to decentralise in recent years, and slaughtered livestock are now generally moved directly from farms, ranches, and feedlots to meat-packers. The cattle-slaughtering sector of the industry, in particular, has become concentrated in cattle-raising regions to the west-namely the western Corn Belt and the Great Plains, where beef is shipped to wholesalers and retailers primarily in the form of fresh primal cuts. Hog slaughtering is still carried on chiefly in large plants, where the hogs are processed into numerous cuts and products.

In accordance with the Humane Methods of Slaughter Act of 1978, all livestock are now made insensible before they are killed. For cattle and sheep, a captive bolt, a type of gun designed for stunning, is generally used. Hogs often are immobilised painlessly through gassing. Many parts of the slaughtered animals are shipped for consumption as fresh meat; other parts, especially of the hog, are cured and smoked. The fatty portions are converted into lard and commercial grease by rendering processes. Bones are converted into glue, fertilizer, animal feeds, and other usable products, including pharmaceuticals; hoofs and horns are used or sold for other purposes.

Refrigeration and Transportation

The meat of slaughtered livestock, after thorough chilling, is shipped in refrigerated motor carriers to cities. There the meat is delivered to wholesale distribution centers and grocery centers. Before cold storage of meat was introduced in the U.S., it was customary to ship the living animals to the eastern parts of the country.

Cattle Slaughtering

In recent years labour-saving devices have been widely adopted in the dressing and processing of cattle. A continuous rail system is generally used today, whereby cattle are chained by one leg and hoisted to a movable pulley on an overhead rail and are then slaughtered; from this position blood leaves the body so quickly that death occurs almost instantly, and for purposes of kosher ritual this method of slaughtering is defined as humane even without prior stunning. The carcasses then move slowly alng a continuously moving rail to stations where each required process is completed, including skinning, disemboweling, and beheading.

Before going into refrigeration, the carcasses are cut down the backbone and split into sides. Beef is shipped to wholesalers and retailers as wholesale cuts, such as chucks, rounds, and loins; in fast-distribution plants the carcass is reduced all the way down to retail consumer cuts. Sides and quarters of beef are still sold, but they constitute less than 5 percent of sales.

Hog Slaughtering

Traditionally, hog slaughtering has involved more complete processing at the packing plant than cattle slaughtering.

The carcasses of slaughtered hogs also move on continuous rail or chain systems. Most often the hog carcass is first conveyed through scalding vats to dehairing machines. Hogs generally are not beheaded or dismembered in any way during the slaughtering and dressing process, which includes eviscerating, washing, and trimming.

Later, however, the carcasses are usually carved up into such cuts as loins, legs, and picnic hams or shoulders. Certain cuts, including loins, are sold fresh, without processing, but most cuts go through one or more processing operations. Although some plants still soak particular cuts in barrels filled with brine, it has generally been necessary to shorten the curing process. To prepare hams, a curing solution is usually pumped internally. Hog bellies, which are used for bacon, are mechanically saturated with curing solution through hollow needles. Many cuts are thoroughly or lightly smoked at the packing plant. Nearly 70 percent of all hogs are slaughtered and fabricated into cuts at the packing plant and shipped to commercial processors who produce a complete line of sausage and cured products.

Legislation

Although inspection by the federal government of meat used in the packing industry was provided for in earlier acts of Congress (1890-91, 1895), comprehensive legislation was not introduced until 1906.

By the act of 1906 all cattle, sheep, goats, and hogs became subject to antemortem and postmortem examination when the meat was to be used in interstate or foreign commerce; later the act was extended to include reindeer. By this legislation about 60 percent of the total meat supply of the U.S., was brought under inspection. The Packers and Stockyards Act of 1921 added further control, directed against trust activities. The Wholesome Meat Act of 1967 requires all meat-packers to meet federal inspection requirements, and agreements to that effect have been made between the various states and the federal government.

ELECTRONIC TONGUE TECHNOLOGY AND SMART PACKAGING

Today, meat packaging and monitoring are a major focus of meat industry-related nanotech R&D. Packaging that incorporates nano materials can be "smart," which means that it can respond to environmental conditions or repair itself or alert a consumer to contamination and/or the presence of pathogens. According to industry analysts, the current US market for "active, controlled and smart" packaging for foods and beverages is an estimated $38 billion - and will surpass $54 billion by 2008.

Scientists at Kraft, as well as at Rutgers University and the University of Connecticut, are working on nano-particle films and other packaging with embedded sensors that will detect meat pathogens. Called "electronic tongue" technology, the sensors can detect substances in parts per trillion and would trigger a colour change in the packaging to alert the consumer if a meat has become contaminated or if it has begun to spoil.

While devices capable of detecting meat-borne pathogens could be useful in monitoring the meat supply, sensors and 'smart packaging' will not address the root problems inherent in industrial meat production that result in contaminated meats: these problems are faster meat (dis)assembly lines, increased mechanisation, a shrinking labour force of low-wage workers, fewer inspectors, the lack of corporate and government accountability and the great distances between meat producers, processors and consumers. Just as it has become the consumer's responsibility to make sure that the meat one is consuming has been cooked long enough to ensure that pathogens have been killed. Consumers will soon be expected to act as their own meat inspectors so that industry can continue to trim safety overhead costs and increase profits.

Nanotechnology and Transportation of Gases

Through nano-technology we can increase or decrease gas transmission rate to suit the packaging requirements of the product. Nanotechnology enables designers to alter the structure of the packaging materials on the molecular scale. With different nanostructure, plastics can gain various gas and water vapour permeability to fit the requirements of preserving fruit, vegetable, beverage, wine and other foods. By adding nano particles, people can also produce bottles and packages with more light- and fire-resistance, stronger mechanical and thermal performance and less gas absorption.

Such nano tweaking can increase the shelf life of meats and preserve flavour and colour. SINTEF-Nanotech to create small particles in the film and improve transportation of some gases through the plastic film to pump out dirty air such as carbondioxide has been developed. Dispersing a filler material into nano particles, creates multiple, parallel layer that force gases to flow through the polymer in a "tortous path", forming complex barriers to gases and water vapour

Nanotech Bio-switch in 'Release on Command' Meat Packaging

Researchers in the Netherlands are going one step further. They want to develop intelligent packaging that will release a preservative if the meat within begins to spoil. This "release on command" preservative packaging is operated by means of a bio-switch developed through nanotechnology to extend the shelf life of meat/ food. Research project in Finland is working on printable indicator, which contains reactive substance that signals if oxygen is present in package.

Meat Packaging Sensors in Defence and Security Applications

Developing small sensors to detect meat-borne pathogens will not just extend the reach of industrial agriculture and large-scale meat processing. The US military however views it a, it's a national security priority. With present technologies, testing for microbial meat-contamination takes two to seven days and the sensors that have been developed to date are too big to be transported easily. Several groups of researchers in the US are developing biosensors that can detect pathogens quickly and easily, reasoning that "super sensors" would play a crucial role in the event of a terrorist attack on the food supply. Researchers at Purdue University are working to produce a hand-held sensor capable of detecting a specific bacterium instantaneously from any sample. They've created a start-up company called BioVitesse. Environmental friendly, lightweight-packaging materials can be made for use in army rations.

Overall Advantages of Nanotechnology on Meat Packaging

- Thermal Stability and Chemical Stability
- Gas, Oxygen, water, etc., barrier.
- Dimensional stability and Flexibility
- Functionality - Anti-counterfeit, Anti-tamper, Anti-microbial, Sensors (temperature, moisture, light, decay)
- High mechanical Strength
- Recyclability.

Companies Working on Nano Meat Packaging

- *CONSTAR® International:* "Diamond Clear" - nanomaterial blended with polyethylene terephthalate - oxygen scavenger
- *Air product polymer:* "AIR FLEX EF 9100 EMULSION" - prevent penetration of liquid or gases and environment friendly.
- *KODAK®:* Antimicrobial food packaging
- Amcol International Co-op - NANOCOR®: Polymer blended with nanocrystal
- *Honey well®:* Polymer nanoclay - gas barrier
- *OX:* Oxygen scavenger
- *HFX:* Hot Filled Application
- *CSD:* Carbonated soft drink
- *SOLPHAS®:* Environment friendly nano-particle for coating plastic film, Antimicrobial packaging
- *BAYON®:* Supply nano-composite nylon film to juice carton
- *KRAFT®:* Working on nano-particle films that willl detect meat pathogens.

Global Overview of Nanotechnology Food

A handful of food and nutrition products containing invisible nano-scale additives are already commercially available. Hundreds of companies are conducting research and development (R&D) on the use of nanotech to engineer, process, package and deliver meat and nutrients to our shopping baskets and our plates. Among them are giant food and beverage corporations, as well as tiny nanotech start-ups.

The U.S., is leading in nanotechnology research, with more than 400 research centers and companies involved and more than $3.4 billion in funding. Europe has more than 175 companies and organisations involved in nanoscience research, with $1.7 billion in funding, and Japan has more than 100 companies working with nanotechnologies. Globally, the market for nanocomposites is expected to grow to $250 million by 2008, with annual growth rates projected to be 18 percent to 25 percent per year. A 2004 report produced by Helmut Kaiser Consultancy, "Nanotechnology in Food and Food Processing Industry Worldwide," predicts that the nanofood market will surge from $2.6 billion today to $7 billion in 2006 and to $20.4 billion in 2010.

Status in India

To promote research and development in nano-science the Government of India has launched a programme "Nano-Material Science And Technology Initiative" (NSTI) in tenth five-year plan One hundred crore rupees have been allocated. Yet, this technology is in budding stage and not entered the field of meat science and technology. Mumbai based "Yash Nanotech" is a business information provider and consultant to entrepreneurs and is aiming to become leading suppliers of nanotech tools, products and services in India.

Hazards and Risks

Although the hazards and risks of nano-technology are not known but can be assumed that like any other technology this may also have associated hazards and risks. Environmentalists are afraid that nano-technology may produce contaminants, which because of their nano-size, may pose to be ultra-hazardous. Even if these particles are not harmful, their interaction with products may be harmful. However, judicious use of nano-technology will find tremendous applications in various fields. Research has shown that nano-sized particles accumulate in the nasal cavities, lungs and brains of rats, and that carbon nano-materials known as 'buckyballs' induce brain damage in fish. Vyvyan Howard, a toxicologist at the University of Liverpool in the United Kingdom, has warned that the small size of nanoparticles could render them toxic, and is of the opinion that full hazard assessments are needed before manufacturing is licensed.

Many interested parties, including the Canadian ETC Group and the insurance company Swiss Re, have expressed their concern over releasing tiny particles, which, because of their small size, are able to travel very far into the environment. They warn that we do not yet know how these particles will act in the environment or what chemical reactions they will trigger on meeting other particles. However, these same groups also concur with nanotechnology advocates who feel the field may offer 'cleaner' technologies, and, ultimately, a cleaner environment. But mostly, the concern is for the lack of research into nanotechnology's potential threats to human health, society and the environment. After witnessing widespread rejection of genetically modified meats, the meat industry may be especially skittish about owning up to R&D on "atomically modified" meat products.

TYPES OF PACKAGING

Practically all films used for meat packaging derive from synthetic "plastic" materials.

Cellulose, which is not a synthetic but a natural material derived from wood, was formerly widely used in the form of transparent films. It is now no longer of great importance in meat packaging although still used for some specific purposes. However, cellulose is still important for the manufacture of certain kinds of artificial sausage casings.

Single-Layer Films

One common use of single-layer films is the wrapping of meat pieces, processed meat products, bone-in or boneless meat cuts or even entire carcasses. These films are usually self-adhesive, *i.e.*, they cling together -"cling film"- in the overlapping areas. Hence they provide good protection from external contamination and to some extend from evaporation, but no protection from oxygen, as they are not hermetically closed or sealed packages. Foils with good self-adhesive properties are PE, PA, PVC and PP.

Another important utilisation for single-layer films is in freezer storage. For meat blocks, meat cuts or smaller portions of meat or meat products, single-layer films are stretched tightly around the meat surface before freezing. The tight film prevents evaporation losses, which occur during freezer storage of unpacked products. The film is in tight contact with the products surface, in order to avoid evaporation, ice formation and freezer burn at non- contact spots. Suitable cold resistant films for freezer storage are PA or PE.

One specific utilisation for single-layer films is the wrapping of chilled meat portions for self-service outlets (supermarkets, etc.). Those meat pieces (beef, pork or chicken) are placed in a hygienic cellulose or plastic tray and tightly wrapped with single-layer plastic film. The ends of the foil are overlapped at the bottom side of the tray, where they firmly cling together. Films to be used should have low water vapour permeability to avoid the drying out of the meat during storage.

But for making attractive to customers, such meat needs to retain an attractive bright red meat surface colour (oximyoglobin) especially in the case of fresh portioned beef. For this reason the plastic foils to be used shall have a high oxygen permeability so that the oxygen of the air can react with the myoglobin of the meat and form the bright red oximyoglobin. Oximyoglobin is not a chemical compound but a loose aggregation of oxygen to the red meat pigment myoglobin, which keeps meat bright red for a number of hours. Suitable single-layer films for fresh meat packaging are PE or soft PVC. Formerly cellulose films were also used, which have the same permeability pattern but are less self-adhesive and the overlapping ends do not cling together very well.

Multi-Layer Films

Practically all the other films used for meat packaging are designed as strong oxygen and water-vapour barriers. In order to fully achieve these requirements, films with good barrier properties for oxygen and water vapour respectively are combined.

A very efficient combination is PA/PE. PA is used as the outside layer for example for films for vacuum bags. PA is relatively oxygen proof but permeable to some extend to water vapour.PE has the opposite properties, it is water vapour proof but permeable to oxygen. The combination of both renders such a multi-layer film very tight against oxygen and water vapour evaporation. Moreover, the PE used as the inside layer has good thermoplastic properties and is therefore well suited for heat sealing. The PA/PE combination is the most simple structure for a multi-layer film. The packaging industry has refined the systems by introducing additional layers which serve as strong oxygen barriers.

Sealant layers consist mostly of Polyethylene (PE) or Ionomer (I). Outside layers may be Polyamide (PA), Polyester (PET) or Polypropylene (PP). Barrier layers for oxygen are made of Polyvinylidenchloride (PVDC) or materials with similar properties.

Vacuum bags, used for vacuum packaging machines are composed of two or more sheets of multi-layer films. By drawing the vacuum and sealing of such bags, the air is excluded from the package and the damaging effects of oxygen such as rancidity or discolouration of the packed products will be significantly slowed down or not develop at all. However, exposure to strong light may cause discolouration even under vacuum.

Processed meat products in slices or as entire pieces are packed in small to medium-size vacuum bags. For larger sized products, bags made of shrinkable films can be used where, after vacuum-packaging, the product in its package of synthetic film is sprayed with or dipped into hot water (80°C).

The contact with the hot water causes the shrinkage of the thermoplastic film and results in tight impermeable wrapping of the goods.

Vacuum packaging for fresh chilled boneless beef cuts is mainly done to promote meat ripening. The beef meat cuts can remain for a number of weeks (maximum 3 months) in the vacuum bag, provided oxygen remains excluded and the storage temperature is kept at -1°C (which is just above the freezing point of meat). During these storage conditions lactic acid bacteria (which do not cause spoilage!) will inhibit the growth of most other bacteria, which results in the prolonged microbial stability. Beef becomes very tender during such extended ripening periods without losing much of its typical flavour. Heat treatment or cooking for some meat products can be carried out in the package after vacuum-packing. Temperatures of 60°-80°C or even higher up to sterilisation temperatures (above +100°C) can be employed for hams, sausages, etc. In these cases a pasteurisation or sterilisation effect of the uncooked packaged products is achieved and re-contamination avoided as long as the package is not opened.

For specific products such as entire sausages, semi-automatic vacuum-packaging can be employed. A bottom film is moulded according to the shape of the sausages by using heat and force (by compressed air or mechanical). These machines are called thermo-formers. The sausages are loaded and a rigid top film is sealed on after evacuating the moulded spaces. Individual product portions are cut apart along their sealing layers.

A packaging method commonly used in larger meat industries is skin packaging. For this method the products are placed in the packaging machine, usually on a rigid film, which serves as the bottom layer of the final package. Another flexible film (top layer, which is heated for increased flexibility) drapes itself from above around the product, resembling a tight "skin" on the product surface avoiding wrinkles and purges. The skin-like coverage of the product takes place in a sealing station in the packaging machine, where the top and bottom film are sealed around the edges. Individual packages are separated by cutting around the bottom seal perimeter.

The latest development in this sector is the "form-shrink" packaging technology. Products, *e.g.*, meat cuts, chicken carcasses, entire sausages, smaller portions of meat products, are placed between two shrinkable films, which are moulded without wrinkles around the goods. Sealing seams can be kept extremely small. This technology is very cost-effective in terms of usage of packaging films but requires high-tech equipment and is only of relevance for large-scale industries.

A useful technology is Modified Atmosphere Packaging (MAP) of meat and meat products. The packaging materials used are gas-proof multi-layer films composed for example of PE, PA and barrier layers. Rigid films may be used to mould cup or box shaped containers which are filled. A flexible lid foil is then sealed on. MAP packaging can also be done for ordinary plastic bags/pouches.

MAP packages are firstly subjected to a vacuum. A mixtures of gases is the introduced into the air-free space before sealing. The gas mixture usually contains nitrogen (N_2) and carbon dioxide (CO_2). N_2, which is also the major constituent of atmospheric air, is inert, *i.e.*, it does not react with meat product components such as fat or myoglobin. Its function is to replace the atmospheric oxygen (O_2) and thus prevents O_2 induced negative impacts. The other component of the gas mix, CO_2, has a protective function, as it inhibits to some extend the growth of bacteria and moulds.

The gas mixture commonly used is 20 per cent-30 per cent CO_2 and 70 per cent-80 per cent N_2. This is applicable for all processed meat products. If fresh meat pieces are to be packed in gas-proof packages instead of wrapping them with oxygen-permeable foil, the bright-red fresh meat colour can be achieved by adding oxygen to the gas mix to be injected into the package and replacing the N2 content accordingly. As sufficient oxygen is needed to maintain the bright-red colour, gas mixes for fresh meat are usually composed of 70 per cent-80 per cent O_2 and 20- per cent-30 per cent CO_2. Skin- and MAP-packaging are often too sophisticated for the small producers, but may be of increasing interest to medium-size meat plants. There are now small manual and semi-automatic packaging machines available, which are designed for smaller throughputs. However, the utilisation of those machines implies that the necessary types of synthetic films, and in the case of MAP-packaging, also the relevant gases, are available.

USE OF NANOTECHNOLOGY IN MEAT PACKAGING

The flexible packaging industry is growing rapidly. It is a $38 billion global market. With the demand for flexible packaging growing at an average rate of 3.5 percent per year, flexible materials need to meet and exceed the high expectations of consumers and the stressors of the supply chain. Increasing competition between suppliers, along with government regulations, have resulted in innovations in films that enhance product and package performance, as well as address worldwide concerns with packaging waste.

The consumer demands meat, to remain fresh for long time, ease in handling, safe and healthy with environmental friendly packaging. Properties such as mechanical and heat resistance can be increased. Packaging materials that have improved temperature performance can be used for hot fill operations. Very thin films that can offer the advantages of flexibility and functionalities like being anti-counterfeit, anti-tamper and anti-microbial should be made. Self-heating feature can also be incorporated in the packaging material. Environment friendly, lightweight-packaging materials can be made for use in army rations. In future, with the aid of nanocomposites we may be able to modify plastic into a super barrier just as glass or metal.

Nanocomposite Technology

Nanocomposites, defined as polymers bonded with "nano" particles to produce materials with enhanced properties are recently gaining momentum in mainstream commercial packaging use.

One such innovation is polymer nanocomposite technology, which holds the key to future advances in flexible packaging. "Nanocomposites appear capable of approaching the elusive goal of converting plastic into a super barrier - the equivalent of glass or metal-without upsetting regulators.".

Polymer nanocomposites are constructed by dispersing a filler material into nanoparticles that form flat platelets. These platelets are then distributed into a polymer matrix, creating multiple, parallel layers that force gases to flow through the polymer in a "tortuous path," forming complex barriers to gases and water vapour. The more is tortuosity present in a polymer structure, the higher is the barrier properties. The permeability coefficient of polymer films is determined using two factors: diffusion and solubility coefficients, *i.e.*, P = DxS.

Different types of fillers are utilised; the most common is a nanoclay material called montmorillonite - layered smectite clay. Clays in a natural state are hydrophilic, while polymers are hydrophobic.

To make the two compatible, the clays' polarity must be modified to be more "organic." One way to modify clay is by exchanging organic ammonium cations [positively charged ions] for inorganic cations from the clays' surface. Many different types of commercial plastics, flexible and rigid, are utilised for nanocomposite structures, including polypropylene, nylon, polyethylene terephthalate (PET) and polyethylene.

Chemical giant Bayer produces a transparent plastic film (called Durethan) containing nanoparticles of clay. The nanoparticles are dispersed throughout the plastic and are able to block oxygen, carbon dioxide and moisture from reaching fresh meats or other foods. The nanoclay also makes the plastic lighter, stronger and more heat-resistant.

Nylon nanocomposites, used as barrier layers for multilayer PET containers, prove to perform as much as two to three times better than the traditional ethylene vinyl alcohol barrier layers, since nylon has a 50°F higher melt temperature.

The advantages of nanocomposite films are numerous, and the possibilities for application in the packaging industry are virtually endless. Because of the nanocomposite process's dispersion patterns, the platelets result in largely improved performance in the following properties:

- Gas, oxygen, water, etc., barriers
- High mechanical strength
- Thermal stability
- Chemical stability
- Recyclability
- Dimensional stability
- Heat-resistance

Good optical clarity (since particles are nano-size). By 2009, it is estimated that the flexible and rigid packaging industry will use 5 million lb of nanocomposite materials in the food and beverage industry. By 2011, consumption is estimated to be 100 million lb. Polymer nanocomposites are the future for the global packaging industry.

Once production and materials costs decrease, companies will be using this technology to increase their product's stability and survivability through the supply chain to deliver higher quality to their customers while saving money. The advantages that nanocomposites offer far outweigh the costs and concerns, and with time, the technology will be further refined and processes more developed. Other types of nanofillers, allowing new nanocomposite structures with different, improved properties that will further advance nanocomposite are being developed for uses in many diverse packaging applications.

Antimicrobial and Active Packaging using Nanotechnology

Nano materials helps to keep products fresh for a longer period of time by using nano-sensors placed in meat production and distribution facilities, meat packaging or the meat itself which can detect all kinds of meat borne pathogens like E. coli, Campylobacter, Salmonella and Listeria spp by attaching themselves to the pathogens. Nanostructured film can prevent the invasion of bacteria and microorganisms and ensure the meat safety, a consultancy said. With embedded nanosensors in the packaging, consumers will be able to determine whether meat has gone bad or find out its nutrition content.

Anti-microbial packaging materials are very useful. It is hoped that with the advancement and growth in nano-science we can expect almost all packaging materials with anti-microbial agents incorporated in them. By placing Nano-sensors in packages, problems such as contamination, toxins or lapses in expiry dates can be detected.

Another window of opportunity for nano-particles is represented by meat safety applications. Latour *et al.* (2003) are investigating the ability of synthesised adhesin-specific nano-particles to irreversibly bind to targeted types of bacteria, inhibiting them from binding to and infecting their host. This research is aimed at reducing the infective capability of human meat borne entero-pathogens in poultry products, using two types of nano-particles. One type is based on the self-assembly of organic polymer (*e.g.*, polystyrene), and the other on inorganic nano-particles functionalised with polysaccharides and polypeptides that promote the adhesion of the targeted bacterial cells. Understanding the interaction between contaminated surfaces and microorganisms allowed the design of materials that are resistant to bacterial adhesion. Researchers are already making efforts to develop a new generation of "self-cleaning" materials loaded with anti-microbial compounds that can be released under certain environmental conditions and kill the contaminant micro flora.

Baker and coworkers have used high-shear mixing of a lipid oil discontinuous phase with an aqueous continuous phase to develop nano-emulsions, which is an effective solution for disinfecting sensitive equipment safely. Nano-emulsions consist of oil droplets of 400-800 μm in diameter that are able to fuse with and subsequently disrupt the membrane of a variety of different pathogens, such as bacteria, spores, enveloped viruses, and fungal spores.

Kodak, best known for producing camera film, is using nanotech to develop antimicrobial packaging for meat products are commercially available since 2005. Kodak is also developing other 'active packaging,' which absorbs oxygen, thereby keeping meat fresh.

THE LATEST TRENDS IN MAP AND OVERWRAP PACKAGING

Usually discussions about packaging technology center on new advancements that have arrived or will arrive soon. Features are unveiled, and their unique processing benefits checked off and dissected. Less often, but perhaps more important, the conversation takes a different direction, to how packaging truly aids retail merchandising and addresses things important to shoppers.

Looking at modified-atmosphere packaging and overwrap packaging, little change has occurred in recent times. Broadly speaking, overwrap packaging is still the most common format in retail meat cases. Of course, much of that use comes from in-store packaging. In fact, according to most surveys, its dominance was slowly trending downward until a few years back, when some retailers of significant size and influence decided that the in-store wrapped look for certain key beef cuts was an important selling point for shoppers.

Their switch from case-ready, modified-atmosphere packages to PVC overwrapped packages distributed in mother bags stabilised the decline. Perhaps it is a sign of the times, but modified-atmosphere packaging has not experienced much growth either over the past two or three years. An experienced observer of the packaging and retail industries senses that MAP may have seen a slight uptick, with some gains in ground beef, but otherwise it has been basically flat. It remains the third most-common packaging format in the meat case after overwrap and vacuum packaging.

There does not appear to be much to get excited over here. But circling back to shoppers and their interests, MAP seems to soothe a couple of important pains. According to a packaging industry source, the longer overwrap packages are in distribution or in the display case, meat purge tends to work its way over the tray lip and under the package.

Shoppers avoid leaky packages and those that appear leaky. They do not like how they look, and they certainly do not want to handle them. The hermetic seals of modified-atmosphere packages, if done properly, keep the packages clean and dry. Cross-contamination concerns are eliminated for shoppers and retailers, alike, and fewer products must be marked down, reground or thrown out.

But overwrap packaging's decades-long inertia continues on because it works. It is what has always been done, and moving away from it at anything more than a measured pace means change. And change is hard.

Successful change requires commitment to more than just switching packages. It means implementing education programmes not only for shoppers, but for store personnel as well. The market is up to doing this. But realistically speaking, retailers have so many priorities to worry about, that if a package is not causing pain and it works, then change will happen when it does.

8

Principles of Meat Drying

Drying meat under natural temperatures, humidity and circulation of the air, including direct influence of sun rays, is the oldest method of meat preservation. It consists of a gradual dehydration of pieces of meat cut to a specific uniform shape that permits the equal and simultaneous drying of whole batches of meat. Warm, dry air of a low humidity of about 30 percent and relatively small temperature differences between day and night are optimal conditions for meat drying. However, meat drying can also be carried out with good results under less favourable circumstances when basic hygienic and technological rules are observed. Intensity and duration of the drying process depend on air temperature, humidity and air circulation. Drying will be faster under high temperatures, low humidity and intensive air circulation.

Reducing the moisture content of the meat is achieved by evaporation of water from the peripheral zone of the meat to the surrounding air and the continuous migration of water from the deeper meat layers to the peripheral zone.

There is a relatively high evaporation of water out of the meat during the first day of drying, after which it decreases continuously. After drying the meat for three or four days, weight losses of up to 60-70 percent can be observed, equivalent to the amount of water evaporated. Consequently, moisture losses can be monitored by controlling the weight of a batch during drying. Continuous evaporation and weight losses during drying cause changes in the shape of the meat through shrinkage of the muscle and connective tissue. The meat pieces become smaller, thinner and to some degree wrinkled. The consistency also changes from soft to firm to hard.

In addition to these physical changes, there are also certain specific biochemical reactions with a strong impact on the organoleptic characteristics of the product. Meat used for drying in developing countries is usually derived from unchilled carcasses, and rapid ripening processes occur during the first stage of drying as the meat temperature continues to remain relatively high. For that reason the specific flavour of dried meat is completely different from the characteristic flavour of fresh meat. Slight oxidation of the meat fats contributes to the typical flavour of dried meat.

Undersirable alterations may occur in dried meat when there is a high percentage of fatty tissue in the raw meat. The rather high temperatures during meat drying and storage cause intensive oxidation (rancidity) of the fat and an unpleasant rancid flavour which strongly influences the palatability of the product.

Meat drying is a complex process with many important steps, starting from the slaughtering of the animal, carcass trimming, selection of the raw material, proper cutting and pre-treatment of the pieces to be dried and proper arrangement of drying facilities. In addition, the influence of unfavourable weather conditions must also be considered to avoid quality problems or production losses. The secret of correct meat drying lies in maintaining a balance between water evaporation on the meat surface and migration of water from the deeper layers.

In other words, care must be taken that meat surfaces do not become too dry while there is still a high moisture content inside the meat pieces. Dry surfaces inhibit the further evaporation of moisture, which may result in products not uniformly dried and in microbiological spoilage starting from the areas where the moisture content remains too high.

Adherence to the following meat-drying techniques should avoid failures in the production of dried meat and ensure obtaining products of good quality with a long shelf-life. The following description of the basic technology of meat drying includes the salting of the meat before drying. Presalting is not absolutely necessary, but has certain advantages, particularly for the drying of meat strips and large flat meat pieces and is therefore strongly recommended for this type of product.

PARAMETERS OF SOLAR DRYING OF FRESH MEAT

The processes which take place during meat drying are mainly *physical*. The aim is the reduction of the moisture content of the meat in warm and relatively dry air in order to achieve low*a_w-values*, where microbial growth is stopped and the meat can be stored over several weeks or months without refrigeration.

In addition to the physical changes in the meat during drying, there are also certain *biochemical* reactions with a strong impact on the sensory characteristics of the product. In many developing countries, meat used for drying is usually derived from unchilled carcasses, and rapid meat ripening processes occur during the first stage of drying, as the meat temperature continues to remain relatively high. For this reason the specific flavour of dried meat is different to that of fresh meat. Slight oxidation of remaining meat fats also contributes to the typical flavour of dried meat.

In contrast to the open air *sun drying, solar drying* takes place in closed systems. Here a specific micro-climate is created, with higher temperature and lower relative humidity than in the outside surrounding air, and also with a reinforced air circulation through convection and tentatively with additional fans. These conditions favour the fast evaporation of a substantial part of the meat moisture. Furthermore, even under partly or fully clouded skies, there is still a certain amount of solar energy absorbed by the solar collectors, which keeps the air humidity low in the system, so that the drying process takes place. In contrast, during open air drying or "sun drying", the drying process will be slowed or stop in cloudy weather conditions.

According to tests carried out in the drying chambers of tunnel driers, the micro-climatic conditions to be achieved in closed drying systems can be summarized as follows:*Maximum temperatures* are generally in the range of +50-55°C during day-time with the strongest solar radiation (usually at noon). During the other periods of the day the temperatures gradually decline in relation to the solar impact to reach minimum values of approx. +30°C in the morning and evening hours. In environments with high natural air humidity, the corresponding *air humidity* pattern inside the dryers are in the range of 60% during the lower temperature phase and 20% during the high temperature phase. In dry semi-arid or arid climates, the humidity values in the dryers will certainly be lower. The *speed of the air stream* circulating into the drying chamber is also related to the intensity of the solar radiation. Air speeds in the range of 0,4 m/sec during lower and up to 0,8 to 1 m/sec during high solar impact are the norm. Air speed may be as low as zero if there is no solar radiation, but additional electric fans would allow for continued intensified air circulation. However, in most cases it is possible to operate closed drying systems without electrical fans.

Drying Techniques using Solar Dryers

Meat drying in solar dryers is usually carried out *on trays*. This method allows a higher load capacity in the drying chamber and is less labour intensive than drying *by suspending* the individual meat pieces. Drying of other food like fruits or vegetables in solar dryers is also done on trays. Trays should be built in a way that they can be piled on top of each other with sufficient space for air circulation. They are usually made of wire or fibre mesh for good air penetration and stabilized by a frame.

The simplest technique is to continue the drying of the flat meat pieces until they reach a hard solid texture. The necessary total drying time is approx. *48 hours* including night time. The effective drying periods last for approx. *8 hours* a day during the main solar impact. Closed solar drying systems which are protected against external influences have the advantage of leaving the meat in the dryer during the night.

The main dehydration effect takes place in the first phase of the meat drying, hereafter dehydration decreases continuously. Depending on the type of meat (high or low water binding capacity) and the size of the meat pieces (larger or smaller, flat or strips), arrives after one day's meat in the solar dryer reduces to *45-35%* of its original weight and after two days at *30-20%*.

This corresponds to a remaining moisture contents in the meat of *40-45%* after one day and *12-18%* after two days respectively. The water activity of such fully dried meat lays in the range of *0.5-0.6* and is low enough to inhibit bacterial growth (for microbial growth a_w 0.75 or above is required).

Quality of the Finished Product

Drying of fresh untreated meat of the shape described (strips or flat) takes at least two days, in many cases three to four days.

After this period the dried meat is ready for consumption and can be packaged, stored and/or transported. At this stage the product should meet the following quality criteria:

The appearance of the dried meat should be as uniform as possible. The absence of large wrinkles and notches indicates the desired steady and uniform dehydration of meat.

The colour of the surface, as well as of the cross-cut should be uniform and dark red. A darker peripheral layer and bright red colour in the centre indicates excessively fast drying. Because of the remaining higher water content in the centre, these meat parts may still be susceptible to microbiological growth.

The texture of properly dried meat must be hard, similar to frozen meat. A softer texture can be recognized by pressing the meat between your fingers. These pieces should be kept for one more day in the dryer for finishing.

Taste and flavour are very important criteria for the acceptance of dried meat by the consumer. Dried meat should possess a mild salty taste which is characteristic for naturally dried meat with no added spices. Off-odours must not occur. However, a slightly rancid flavour, which occurs because of chemical changes during drying and storage, is commonly found in dried meat and is acceptable. Dried meat with a high fat content should not be stored for a long period but used as soon as possible in order to avoid intensive rancidity.

PREPARATION OF DRIED MEAT FOR CONSUMPTION

After completion of the drying process, the dry meat is normally packaged, preferably in moisture-proof plastic bags to avoid absorption of moisture during storage. For consumption, the dried meat needs to be rehydrated by submerging it in water or it can be directly added to meals.

Instead of rehydrating entire pieces the dried meat can be chopped into smaller pieces or even used as dry flakes or similar particles. The traditional way to reduce the size of the meat pieces is to pound it with a wooden stick in a wooden bowl. The resulting coarse dry powder is commonly used for food preparations such as soups, meals with gravy, etc.

The comminution of dried meat can also be included during the drying process in order to transfer such meat into easier-to-use food. One useful method is the semi-drying of meat pieces for only 12 hours in the tunnel dryer so that the meat pieces become semi-rigid. Although they still have a relatively high moisture content, this does not render them shelf-stable. In a second phase, these semi-rigid meat pieces are passed through a grinder (5-6 mm plate) and the minced particles exposed for another day to solar drying. The resulting dry minced pellet-like meat particles can easily be packaged in plastic bags and stored without the risk of absorbing moisture. Rehydration of these pellets is easy - just add water (1:1 – 3:1, rehydration time 30-60 min). In the rehydrated form, the meat can be used for the fabrication of hamburger patties, meat balls etc. In dry form, it can be added to soups or meals as desired.

TYPES OF MEAT SUITABLE FOR DRYING

Meat drying is a simple but efficient food preservation activity. Dried meat can be stored under ambient temperatures for many months. Due to the low water content, microbial spoilage of the muscle proteins can be safely prevented. However, deterioration of adhering fatty tissue through rancidity cannot be stopped. It is therefore advisable to use lean meat only. *Beef* and*buffalo meat* as well as *goat* and certain *game meats* (deer, antelopes) are best suited. The same applies to meat of livestock used in some regions for meat production, such as*camels* or *yaks*. The suitability of *mutton* is ranked slightly lower. *Pork*, even from very lean muscle parts, is less suitable, as it contains higher amounts of intermuscular and mostly invisible intramuscular (within the muscle cells) fat, which is prone to oxidation and hence turns quickly rancid.

Preparation of Meat for Drying

The meat is exposed to the open air and intermittent solar radiation and quickly loses substantial amounts of its tissue moisture. The drying process will be faster the shorter the distance from the centre of the meat piece to its surface. In order to accelerate the drying process in particular from the inner layers of the meat, it is therefore common practice to cut the meat in*narrow strips* or in *flat pieces*.

Recommended shapes for meat pieces to be dried are:

- strips with a rectangular cross-section of 1 x 1 cm
- flat- or leaf-shaped pieces with cross-sections of 0,5cm x 3 to 5cm.

In large thick meat pieces, the moisture content in the centre would remain high for too long and, given the high ambient temperatures, could easily lead to microbial *spoilage*, as microorganisms still would find good conditions for growth. Hence flat meat pieces should always be used for successful drying. Spoilage through chemical reactions can occur when fat turns rancid. Adhering visible fatty tissues need therefore to be carefully trimmed off from the lean meat in order not to limit the shelf life of dried meat.

Meat Drying Techniques

For the traditional drying of meat, the natural conditions *sunshine* and *air circulation* are used. Two drying techniques can be distinguished with both using prevailing natural conditions but differing in the impact of the solar energy. These techniques are called *sun drying* and *solar drying* and are described hereunder.

Sun Drying

The basic traditional drying method is called *sun drying,* characterized by direct solar radiation and natural air circulation on the product. Meat pieces are cut into strips or flat leaf-shaped pieces as described above. They are then *suspended* in the open air or *spread* on drying trays made of fibre or wire mesh with a wooden or metallic frame. For sun drying, in particular for the suspension method, the meat is sometimes *dipped in salt solution* (approx. 14% common salt). This helps to limit microbial growth on the meat surfaces and protects to some extend against insects.

The sun drying method is known to have certain disadvantages, such as exposure to *contamination* from sources such as dirt, wind, rain, insects, rodents and birds. Quality deficiencies, such as changes in colour, off-flavours, foreign contaminating substances such as dirt and sand and even high surface microbial contamination may occur. Heavy microbial contamination can affect the meat after rehydration, when sufficient moisture for renewed bacterial growth is available, as this will lead to product deterioration and even possible food poisoning.

The *sun drying* method can be done at the domestic or farm level for quick and uncomplicated preservation for example of surplus meat which cannot be consumed immediately or stored properly. Possible contamination can be limited at the small-scale level, as these operations are easy to manage and supervise. For relatively small pieces of meat the drying process can be completed in the course of one day by starting in the morning. If the drying cannot be finished in the lapse of 8-10 hours, which is usually the case with meat pieces of a thickness >1 cm, the meat should be stored overnight in a safe and dry place and exposed the following morning again to the sun.

Solar Drying

For larger-scale or commercial meat drying operations in rural settings, improved approaches have been developed using solar drying. In contrast to sun drying, where the meat is exposed directly to the sun, the solar drying method uses indirect solar radiation. The principle of the solar drying technique is to collect solar energy by heating-up the air volume in *solar collectors*and conduct the hot air from the collector to an attached enclosure, the *meat drying chamber*. Here the products to be dried are laid out.

In this closed system, consisting of a solar collector and a meat drying chamber, without direct exposure of the meat to the environment, meat drying is more hygienic as there is no secondary contamination of the products through rain, dust, insects, rodents or birds. The products are dried by hot air only. There is no direct impact of solar radiation (sunshine) on the product. The solar energy produces hot air in the solar collectors. Increasing the temperature in a given volume of air decreases the relative air humidity and increases the water absorption capacity of the air. A steady stream of hot air into the drying chamber circulating through and over the meat pieces results in continuous and efficient dehydration.

Construction of Solar Dryers

Solar dryers are composed of two essential parts, the *solar collector* and the *drying chamber*. Both are simple constructions and can be built everywhere using locally available materials such as timber, brick, metal plates and transparent plastic sheets. Two examples of solar dryer design – *tunnel dryer* and *multi-collector dryer* - are provided hereunder.

In the *tunnel dryer*, the shape of the solar collector is tunnel-like to provide enough surface area for the absorption of the solar radiation. For the capacity of the drying chamber of 50-100 kg of goods the length of the collector should be 10-15 meters.

The *multi-collector dryer* uses a collector surface of approximately the same area as the tunnel dryer, but subdivided into three individual collectors

. The three collectors are positioned in a way that one collector points to the morning sun, another to the mid-day sun and the last to the afternoon sun. Hence the solar radiation is utilized in an optimal way. Moreover, the angled positioning of the collectors enables the easy conduct of the hot air, which has the tendency to raise up, into the raised drying chamber.

The top parts and preferably also the side parts of solar collectors must be permeable for solar radiation. These parts need to be of transparent material (uv-resistant plastic foil or glass). The bottom part should be of metal plate, possibly corrugated to increase the surface, and black in colour for optimum absorption of solar heat. It is recommended to insulate the bottom part of the collector underneath. A sophisticated approach would be to use a black metal coated polyurethane panel, but a similar effect can be achieved by insulating the bottom plate with low cost insulation material such as styrofoam.

The major part of the thermal energy absorbed is inside the collector, as it is permeable for solar radiation, and the heated air remains trapped inside the system. The only outlet for the hot air is the opening leading to the drying chamber, where the hot air is continuously moved by natural convection only. This movement can also be supported by fans. In order to keep the air stream going continuously, there must be an air inlet protected by wire mesh at the front side of the collector part of the dryer.

An important feature of solar drying devices is the *size of the solar collectors*. Depending on the quantity of goods to be dried, collectors must have the capacity to provide sufficient quantities of hot air to the drying chamber. Collectors which are too small in proportion to the amount of meat to be dried will result in failed attempts and spoiled meat.

In case of the *tunnel dryer*, the collector can be built horizontally or with a slight slope towards the higher positioned drying chamber for easy convection of the heated air from the collector to the chamber. For horizontal tunnel dryers, the air movement can be supported by electric fans attached to the front part of the dryer. In rural areas, where no electricity is available, good results can be achieved by using photo-voltaic panels to drive the fans. But in most cases, fans may not be necessary at all, as by natural convection, enough hot air moves into the drying chamber.

The drying chamber is a closed wooden or metal-sheet construction, which has an air inlet from the collector side and a door which can be used for loading/unloading. At the rear in the upper section of the chamber, there must be an opening, which serves as the air outlet for the hot air conducted into the chamber. The opening must be protected by wire mesh to avoid the access of insects, rodents and birds. Inside the drying chamber there are on several levels of horizontal drying trays consisting of frames with wire mesh to hold the meat pieces to be dried. Alternatively, for drying of vertically suspended meat pieces, hanging sticks may be used.

SELECTION OF MEAT FOR DRYING

As a general rule only lean meat is suitable for drying. Visible fatty tissues adhering to muscle tissue have a detrimental effect on the quality of the final product. Under processing and storage conditions for dry meat, rancidity quickly develops, resulting in flavour deterioration. Dry meat is generally manufactured from bovine meat although meat from cameloids, sheep, goats and venison (*e.g.*, antilopes, deer) is also used. The meat best suited for drying is the meat of a medium-aged animal, in good condition, but not fat. Meat from animals in less good nutritional condition can also be used for drying, but the higher amount of connective tissue is likely to increase toughness.

It is very important that raw material for the manufacture of dry meat is examined carefully for undersirable alterations such as discolouration, haemorrhagic spots, off-odours, manifestation of parasites, etc. Such defects must be trimmed off.

Carcasses have to be properly cut to obtain meat suitable for drying. Owing to their size, beef carcasses are more difficult to handle under rural conditions than carcasses of sheep, goats or game. In the absence of chilling facilities, beef carcasses must be cut and deboned immediately after slaughter.

Beef Carcass Cutting, Trimming and Deboning

Carcass Cutting

The carcass is first split into two sides along the spinal column and then cut into quarters. Fore- and hindquarters are separated after the last rib, thus leaving no ribs in the hindquarter. For suspension the hindquarter is hooked by the Achilles tendon and the forequarter by the last two ribs.

Trimming

After the quarters are suspended so that they do not touch the floor or anything around them, they are trimmed. Careful trimming is very important for the quality and shelf-life of the final product. The first step is to remove with a knife all visible contamination and dirty spots. Washing these areas will spread bacterial contamination to other parts of the meat surface without cleaning the meat. After completing the necessary cleaning of the meat surfaces, knives and hands of personnel must be washed thoroughly. Using a sharpened knife, the covering fat from the external and internal sides of the carcass and the visible connective tissue, such as the big tendons and superficial fasciae, are carefully trimmed off.

Deboning

It is recommended that this operation should start with the hindquarters and follow with the forequarters. The aim is to remove the bones with the least possible damage to the muscles. Incisions into the muscles are inevitable but only at spots where the bones adhere and have to be cut off.

Deboning of the suspended hindquarter should start from the leg and proceed to the rump and muscles along the vertebral column.

Deboning of the forequarter must start with cutting and deboning the shoulder separately, followed by cutting off the rib set, together with the intercostal muscles.

Deboning of the forequarter is completed by removing the meat from the neck and the breast region of the spinal column.

TYPICAL DRIED-MEAT PRODUCTS

Some examples of dried meat or dried and further processed meat manufactured in Africa, America and the Near East are described. No mention is made of the various dehydrated meat products well known in the Far East. These products are somewhat different owing to their sugar component.

Odka

Odka is basically a sun-dried meat product made of lean beef and is of major importance to nomads in Somalia. In the face of perennial incidence of drought in the Horn of Africa, odka has become important since it is often prepared from drought-stricken livestock.

The production of odka is similar to the simple drying technique described earlier. However, the meat strips cut for drying are bigger and dry salting is usually applied instead of brine salting. After only four to six hours' sun-drying the large pieces of meat are cut into smaller strips and cooked in oil. After this heat treatment drying is continued and finally sauces and spices are added. For storage odka is again covered with oil and, when kept in a tightly closed container, it has a shelf-life of more than 12 months.

Qwanta

Qwanta is manufactured from lean muscles of beef which are further sliced into long strips ranging from 20 to 40 cm and are hung over wire in the kitchen to dry for 24 to 36 hours. Prior to drying, the strips are coated with a sauce containing a mixture of salt (25 percent), hot pepper/chilli (50 percent) and aromatic seasoning substances (25 percent). After air drying the meat pieces may be further exposed to a light wood smoke and are then fried in butter fat and dried again to some extent. At this stage the product is ready for consumption or storage.

Kilishi

Kilishi is a product obtained from sliced lean muscles of beef, goat meat or lamb and is made on a large scale under the hot and dry weather conditions prevailing from February to May. It is produced by sun-drying thin slices of meat. However, recent experience indicates that kilishi can also be produced industrially using tray-drying in a warm air oven. Connective tissue and adhering fatty material are trimmed off the meat which is cut with a curved knife into thin slices of about 0.5 cm thickness, 15 cm length and as much as 6 cm width.

Traditionally, the slices of meat are spread on papyrus mats on elevated platforms or tables in the sun for drying. However, these papyrus mats may lead to hygienic problems, especially after repeated use. Therefore, easily washable corrosion-free wire nets or plastic nets are recommended for horizontal drying. The vertical drying method is also recommended in this case.

Sun-drying of kilishi could also be improved by the use of solar dryers. These devices will increase the rate of drying of the product and keep insects and dust from the product. In the first stage of drying, which takes two to six hours, the moisture of the meat slices has to be reduced to about 40 to 50 percent. The slices are then put into an infusion containing defatted wet groundnut cake paste or soybean flour as the main component (about 50 percent), and is further composed of water (30 percent), garlic (10 percent), bouillon cubes (5 percent), salt (2 percent) and spices such as pepper, ginger and onion. The "dried" slices of meat should absorb the infusion up to almost three times their weight. After infusion, the wet product is again exposed to the sun to dry. Drying at this stage is much faster than at the first stage. When the moisture content of the slices has been reduced to 20 to 30 percent, a process which takes two to three hours depending on weather conditions and the dimensions of the product, the slices are finally roasted over a glowing fire for about five minutes. The roasting process helps to enhance desirable flavour development and to inactivate contaminating micro-organisms. Roasted kilishi is therefore superior in flavour to the unroasted version.

After roasting, the final moisture content ranges between 10 to 12 percent. It will decrease during storage at room temperature to as low a level as 7 percent. When packaged in hermetically sealed, low density plastic bags the product remains remarkably stable at room temperature for a period of about one year.

Biltong

Biltong is a well-known salted, dried meat prepared from beef or antilope meat. Most muscles in the carcass may be used but the largest are the most suitable. The finest biltong with the best flavour is made from the sirloin strip and the most tender is derived from the fillet.

The meat is cut into long strips (1 to 2 cm thick) and placed in brine, or dry-salted, which is actually the most popular method. Common salt, preferably coarse salt (1 to 2 kg for 50 kg of meat), or salt and pepper are the principal ingredients used, although other ingredients such as sugar, coriander, aniseed, garlic or other spices are included in some mixtures to improve flavour. In most cases nitrate or nitrite is added to achieve a red colour and the typical flavour of cured meat. The addition of 0.1 percent potassium sorbate to the raw meat is permitted in South Africa as a preservative. The salt/spice mixture is rubbed into the meat by hand and the salted strips are then transferred to a suitable container. It is recommended that a little vinegar be sprinkled on each firmly packed layer in the container.

Biltong is left in the curing brine for several hours, but not longer than 12 hours (otherwise it will be too salty), and then dipped into a mixture of hot water and vinegar (approx. 10:1). The biltong is now ready for sun-drying for one day. Then the strips are moved into the shade for the rest of the drying period. The product is usually not smoked, but if it is smoked only light cold smoking is recommended, which takes one to two weeks under sufficient air circulation. The biltong is ready when the inside is soft, moist and red in colour, with a hard brown outer layer.

Biltong is sold in sticks or slices. The usual shelf-life is several months without refrigeration and packaging, but in airtight packages the product stores well for more than one year. Biltong is not heated during processing or before consumption. It is eaten raw and considered a delicacy.

Pastirma

Pastirma is salted and dried beef from not too young animals. In some areas camel meat is also used. The meat is taken from the hindquarters and is cut into 50 to 60 cm long strips with a diameter of not more than 5 cm. The strips are rubbed and covered with salt and nitrate. The dosage of the nitrate in relation to the meat is 0.02 percent, that means 2 g of nitrate for 10 kg of meat. Several incisions are made in the meat to facilitate salt penetration.

The salted meat strips are arranged in piles about 1 m high and kept for one day at room temperature. They are turned over, salted again, and stored in piles for another day. Thereafter the meat strips are washed and air-dried for two to three days in summer and for 15 to 20 days in winter. After drying the strips are piled up again to a height of 30 cm and pressed with heavy weights (approx. 1 tonne) for 12 hours. After another drying period of two to three days the meat pieces are again pressed for 12 hours. Finally the meat is again air-dried for 5 to 10 days.

After the salting and drying process, the entire surface of the meat is covered with a layer (3 to 5 mm thick) of a paste called cemen, which consists of 35 percent freshly ground garlic, 20 percent helba (*i.e.*, ground trefoil seed), 6 percent hot red paprika, 2 percent mustard, and 37 percent water. Helba is used as a binder of the paste; the other ingredients are spices, but garlic is the most important as it is antimycotic. The meat strips covered with cemen are stored in piles for one day, and thereafter are dried for 5 to 12 days in a room with good air ventilation, after which the pastirma is ready for sale. Thus, the production of pastirma requires several weeks. However, not much energy is required since most of the salting and drying is done at room temperature. The final product has an average water activity (aw) of 0.88. The a_w should not fall below 0.85 or the meat will be too dry. The salt content should range between 4.5 and 6.0 percent. The product is mould-free for months at ambient temperature even in summer. Pastirma thus has a better microbiological stability than biltong.

Charque

Charque consists of flat pieces of beef preserved by salting and drying. The fresh, raw meat from the fore- and hindquarters is cut into large pieces of about 5 kg, which should not be more than 5 cm thick. The pieces are submerged in a saturated salt solution for about one hour in barrels or cement vats. On removal from the brine, the meat is laid on slats or racks above the brine tank to drain.

For dry-salting, the flat meat pieces are piled on a sloping, grooved, concrete floor under a roof. To form a pile, salt is spread evenly over the floor about 1 cm high. Then a layer of meat is put on the salt. The meat is covered with another (1 cm) layer of salt followed by adding another layer of meat, and so on until the alternate layers of salt and meat reach a height of about 1 m. The pile is then covered with a few wooden planks and pressed with heavy stones.

After eight hours the pile is restacked so that the top meat goes to the bottom of the pile. The restacking process with fresh layers of salt is repeated every day for five days.

The salted meat is then ready for drying. Before initiating drying, the meat pieces are subjected to rapid washing to remove excess salt adhering to the surface. The meat pieces may also be passed through a pair of wooden rollers or a special press to squeeze out some surplus moisture and flatten the meat slabs. The meat is then spread out on bamboo slats or loosely woven fibre mats in a shed or, in industrial production, exposed to the sun on wooden rails which are oriented north-south, thus permitting an even solar coverage.

Initial drying, directly in the sun, is limited to a maximum period of four to six hours. This period of exposure may be subsequently lengthened to a maximum eight hours. Temperatures in excess of 40°C on the meat surface should be avoided. To ensure even drying over the extended muscle pieces, the meat is placed on the rails during the morning and removed again in the afternoon. The meat pieces are exposed to the sun each day over a period of four to five days. After each period of exposure the pieces are collected, stacked in piles on concrete slabs and covered with an impermeable cloth to protect them against rain and wind and to hold the heat absorbed.

When sufficiently dry, the meat pieces are either sold without prior packaging or wrapped in jute sacks. Plastic sacks are not suitable, because the product still contains a certain proportion of its original moisture content, and this moisture must be allowed to drain freely from the product. Charque keeps for months under ambient room conditions and is resistant to infestation by insects and growth of moulds.

9

Meat Preservation and Canning Technology

MEAT PRESERVATION

Meat preservation (including livestock, game poultry), includes a set treatment processes for preserving nutritious properties, taste, texture and color of raw, partially cooked or cooked meats while keeping them edible and safe to consume.

The invention of such processes can be traced back to antiquity, and although some come from 18th and 19th century research, the latest discoveries concerning meat preservation stem from nuclear physics andbiochemistry.

The preservation of the meat has always been a method to prevent against famine, but has also acquired a political and economic dimension in the growth of the human population. In developed countries, this conservation depends largely upon the food industry rather than individuals.

For lesser-developed countries, the Food and Agriculture Organization considers the absence of meat preservation techniques would present a serious obstacle to the development of viable meat production by resource-poor rural livestock producers.

DRYING

Micro-organisms cannot grow unless there is sufficient moisture available to them and drying meat under conditions of natural temperatures and humidity with circulation of air and the assistance of sunshine is the oldest method of preservation.

The free water in a food product, *i.e.*, excluding the water bound to proteins, is termed the water activity (a_w). Free water is that part that can be removed as water vapour (and is not the same as the total moisture content). "Water activity" is defined as the ratio of water vapour pressure measured in the product to the pressure of a saturated water vapour atmosphere at the same temperature.

The minimum moisture content necessary for bacterial growth varies with the type of organism. The lowest value for normal bacteria is water activity 0.91; for normal yeasts it is 0.88; for normal moulds 0.80; and for salt-tolerant (halophilic) bacteria it is 0.77.So water activity must be reduced below these levels to preserve the food.

Muscle meat of almost any kind can be dried but it is necessary to use lean meat since fat becomes rancid during the drying process. Drying involves the removal of moisture from the outer layers and the migration of moisture from the inside to the outside, so the pieces of food must be thin.

The meat is cut into long thin strips or flat thin pieces and preferably salted, either dry or by dipping into salt solution, to inhibit bacterial growth and to protect from insects.

The pieces are suspended in racks in freely circulating air under hygienic conditions and protected from dirt and dust for the several days required. If the air is warm and of low humidity with relatively small temperature fluctuations between day and night the time needed will be shorter. Slow drying allows deterioration since micro-organisms can multiply in the early stages while the moisture content is still high enough. Another problem arises from the practice in developing countries of using meat from unchilled carcasses and while the temperature is still high the meat ripens rapidly so changing the flavour. At the same time there is some oxidation of the fat so further lowering the quality of the finished product.

There are a number of traditional dried products in various regions. For example biltong in South Africa, which is made from beef or antelope meat cut into strips 1 - 2 cm thick, salted, with the addition of nitrate or nitrite, spiced and dried in air for 1 - 2 weeks.

The outer layer of biltong is hard and brown with a soft, inner, red inside, and is eaten raw. It will keep for a year if stored in airtight packaging.

Typical analysis per 100 g: 11.5 g water, 1.9 g fat, 12.5 g ash, 65 g protein, 1.3 MJ (300 kcal).

Jerked beef or charque is the product used in South America which may be made from beef, llama, sheep, alpaca. The fresh meat is cut into large pieces no more than 5 cm thick, salted, pressed for several days and dried - but it still contains moisture which is allowed to drain freely from the product. It keeps for months at ambient temperatures and is resistant to insect infestation and mould growth.

Pemmican is dried meat that has been powdered or shredded and mixed with fat to form a solid product. Typical analysis per 100 g:- 3 g water, 40 g protein, 45 g fat, 2.4 MJ. Pemmican was almost a routine food taken on earlier expeditions until replaced by modern types of dried meat products.

Other traditional dried products include pastirma (Turkey, Egypt and Armenia), odka (Somalia and other countries of East Africa), qwanta (Ethiopia and East Africa) and kilishi (Nigeria and West Africa). There is a variable loss of vitamins from such products due to the long drying times which can be shortened by the use of modern drying techniques.

Such a procedure is freeze-drying which causes little or no loss of vitamins and results in products which are readily rehydrated and much closer in texture and flavour to fresh meat than the traditional dried product but calls for specialised equipment.

Partial Drying/Intermediate-Moisture Foods

In dried meats the water activity is below levels needed for microbial growth so the product is shelf-stable but there will still be chemical and physical changes due to rancidity and discolouration which call for adequate packaging. Some products such as "dry" sausages and hams cannot be dried adequately without spoiling the product - they are termed "semi-dry" - so it is necessary to combine an incomplete reduction in water activity with other methods such as lowering of pH or the addition of nitrate.

In an attempt to avoid the relatively poor texture and flavour of most dried meat products a modern development is partial drying to a moisture content of 15 to 50 per cent and then reduction of free water to the required low levels by adding humectants such as glycerol, sorbitol or other polyhydric alcohols which combine with the free water so that it cannot be used by the micro-organisms.

The meat is cut into small pieces and treated with a mild salt solution, and the humectant and an antimycotic (anti-mould agent) are added and the meat cooked to 70°C before packaging. It will keep for several months even at 38°C but there are changes in texture, colour and flavour. Products preserved in this way are called intermediate-moisture foods and they are more succulent than dried foods but the humectants spoil the palatability and the process has been limited to animal foods in industrialised communities and for military purposes.

Hanging

Although not a method of preservation, a period of 'hanging' can improve the flavour and texture of meats by giving natural enzymes time to break down tough muscle fibres. The temperature range for hanging is 33 to 40 degrees F.

Fresh meat and poultry will rapidly deteriorate in temperatures above 40 degrees F, so be wary of this. Without a refrigerated room, this can obviously only be done during the cold months of the year. Game should be gutted first, and any scent glands should be removed. Hanging times range from 24 to 48 hours, to even longer for a more tender meat (It must remain under 40 degrees F!)

Meat Freezing

Since the time frame of World War II, freezing has become the most popular way to store meat. It is quick and easy and preserves the nutritional value and flavour. Obviously though, freezing depends upon a supply of electrical power. Freezing meat is best at 0 degrees F, for longest shelf life. Wrap all pieces securely in individual moisture-proof packages to prevent freezer burn. Label each package with they type of meat and the date is was frozen. All meats will begin to deteriorate in the freezer though. Ground meats will be good up to 3 months methods-to-preserve-meat and stakes or roasts will last well up to 6 months.

CANNING OF MEAT

Canning meat is convenient and economical, and is not endangered by power failures. Canned foods keep for a very long time (years). Be aware that the greatest danger in canning is botulism, a severe and often fatal form of food poisoning caused by bacteria that thrive in airless conditions. This is easily and readily prevented by being sure to to follow proper canning recipes, and to heat containers for an extended time at 240 degrees F. This temperature requires a pressure canner with its control set for at least 10 pounds. Again, search out recipes for either raw-packed or hot-packed meat.

Micro-organisms can be completely destroyed by heat (sterilisation) but a sterile product can be readily recontaminated unless it is protected. This is achieved by heating in an air-tight can or bottle, or, more recently, in a heat-resistant or aluminium foil-laminated plastic pouch. Sausages can be filled into retortable synthetic casings sealed with aluminium clips. The procedure is to seal the food into the container and then heat it under pressure in an autoclave (retort) to the required temperature for the required length of time and to cool rapidly to avoid overheating. Overheating results in too soft a consistency and a burnt taste. It is not always possible to destroy all the organisms without excessive heat which would spoil the product so the objective is to destroy the greater proportion of the organisms when the remaining few pose no hazard so long as the container is cooled rapidly and stored below 20-25°C. This condition is termed "commercially sterile". The established standard is equivalent to a reduction in the number of micro-organisms by a factor of 10 to the power of 12 so it is clear that the higher the initial load of organisms the more will survive a standard heat treatment.

The intensity of heat treatment necessary depends on the nature of the product, its pH, and the amount of salt and other curing agents present as well as on the bacterial load. The time required at a given temperature will vary with the rate of heat penetration to the centre and so with the size of the container.

The intensity of heat treatment is defined in physical terms called F-value, which means that the product received heat treatment with the same effect on micro-organisms as exposure to a temperature of 121°C for 1 minute. The standard is based on the time required at a temperature of 121°C to destroy all spores of Clostridium botulinum, the most dangerous of all toxin-producing organisms. This is termed "the botulinum cook" and such treatment destroys practically all spoilage and other organisms. It takes 2.45 minutes at 121°C to destroy all C. botulinum spores; this is an F-value of 2.45. Spores of other organisms are less or more heat-resistant. F-value 1 is the lethal effect on micro-organisms after 1 minute at 121°C; F-value 2 (3,4) is the lethal effect after 2 (3,4) minutes.

At temperatures greater than 121°C a shorter time is needed to achieve the F-value of 1, thus at 130°C the time is 0.13 min. Correspondingly the time is longer at lower temperatures, thus at 115°C the time is 4 min at 105°C it is 40 min.

These conditions apply to foods of low acidity (pH above a value of 5) and medium acidity (pH 4.5 - 5); with more acid foods the spores of micro-organisms are less heatresistant. Meat products are mostly low-acid, while meat and vegetable mixtures are medium -acid. In practice once the F-value has been determined for a batch of food according to the size of the container the heat treatment required to treat subsequent batches is the same. Generally it has been shown that F-value 4 will usually ensure commercial sterility. Larger canned products may require F-values up to 20-25 owing to the longer periods required for heat penetration.

A fully-treated product of this type will keep for up to 4 years at ambient temperatures but even fully-preserved meat can contain a very heat-stable spore former, Clostridium sporogenes, which poses a hazard only when stored under extreme climatic conditions, namely at temperatures above about 40°C. If canned meat is to be stored under such conditions then it must be treated more intensively, F-value 12 or more ("tropical preservation") and then has a shelf life up to 4 years.

Virtually every type of meat product made from chopped, cured meat can be canned, as well as stewed meat, dishes in jelly, soups with meat ingredients, and pastas and sausages in brine. Products such as luncheon meats, liver sausage, blood sausage and jellied products are adversely affected by high temperatures and are "three-quarters preserved" at F-values 0.6 to 0.8. The temperature reached at the centre of the pack is between 108 and 112°C and the product is stable for up to 1 year if stored at temperatures no higher than 15°C.

Cooked preserved products are simply boiled until the central temperature reaches near to 100°C and they can be stored (protected from contamination) for 1 year at temperatures no higher than 10°C.

Smaller size containers are most suitable for meat products because heat penetration is mostly by conduction so larger containers would require severe heat treatment involving overcooking. Large pieces of meat products such as hams, shoulders, etc., are pasteurised. Pasteurisation is a more gentle process intended to destroy only pathogenic organisms and the treatment limits the central temperature to about 80°C (Fvalue almost zero). This destroys only vegetative cells and refrigeration is necessary to prevent germination of spores. Pasteurised products must be stored between 2 and 4°C when they have a shelf life up to 6 months. The temperatures quoted must be reached in the centre of the pack to ensure that the entire contents are adequately heated but protein and fat are poor heat conductors. If there is enough liquid in the can, such as meat cooked in gravy, Frankfurters in brine, or through release of liquid from the meat and liquefaction of the fat, heat can penetrate by convection as well as conduction if the can is rotated during the process. This allows a shorter heating time with less damage to flavour, texture and nutrients, and the outer layer of the food is not overheated. Canning operations must be performed only by fully-trained personnel.

High-Temperature Short-Time Processing (HTST)

Since the effect of heat in speeding up biological reactions (in this instance destruction of micro-organisms) is greater than the acceleration of chemical reactions (in this instance damage to protein and other nutrients) heating to a higher temperature for a shorter time is an effective means of preservation. Sterilisationis achieved in a shorter time with less damage to the product. The process is termed high-temperature short-time heating (HTST) and has been particularly applied to milk but can be applied to meat if there is sufficient liquid present to allow mixing of the contents by rotating the cans in the autoclave. The cans must be cooled immediately after the temperature of sterilisation has been reached to avoid overheating.

Nutritional Damage

Most of the investigations of nutritional changes in canning have involved fullypreserved foods. There are large differences reported in the literature due to differences in raw materials and conditions of processing. Thiamin is the most labile of the vitamins and reported losses range between 20 and 40 per cent. This is heat destruction since any water-soluble nutrient that is leached out of the meat will be retained in the can and is usually consumed together with the meat. Losses of niacin and riboflavin are about 10 per cent, 20 per cent of biotin, 20-30 per cent of pantothenate.

If more accurate figures are required they would need to be determined on the specific procedure in the factory in question on the particular product. In principle losses in canning are somewhat higher than in wet cooking since the temperature is higher, and for the parallel reason less than losses in dry cooking methods such as roasting, grilling and frying where the temperature at the surface can be between 180 and 350°C So far as proteins are concerned there is some small reduction in biological value due to reduced availability of methionine and cystine. Semi-preserved meat products heated to temperatures not exceeding 100°C do not suffer this damage. The heat applied causes partial hydrolysis of the collagen so that tough meat, rich in collagen, is rendered more palatable by being canned.

During storage after canning there can be a loss of thiamin of as much as 30 per cent depending on the length of time, and at higher storage temperatures there can be a reduction in protein quality.

Fat Embedding

A traditional process that is parallel to canning is that of cooking the meat in a vessel that can be sealed under a layer of melted fat and so protected from recontamination. An example of such a product is mixiria of the Amazon region where the meat is roasted, sliced and sealed in jars. The layer of fat not only protects the meat from contamination but excludes oxygen, however, organisms can survive so the method is not dependable.

The process of fat embedding was tested by the Australian meat trade in the 19th century - beef was packed into barrels and covered with fat heated to 150°C - but superseded by refrigeration. It is used to a very limited extent in some industrialised countries, in particular for a product called "potted shrimps", which have been cooked in butter and sealed into jars.

Ionising Irradiation

Micro-organisms can be destroyed by subjecting a food to ionising radiation produced from radioactive or electromagnetic sources. High doses (50 kiloGrays - kGy) are required for sterilisation of meat while recommendations of WHO and legislation in most, if not all, countries limit the dose at present to 10 kGy, simply because safety has been established up to this level. The 10 kGy dose does not sterilise the product but substantially reduces the bacterial load and is effective in destroying many pathogens including Salmonellae. A dose of 2 - 5 kGy will extend the shelf life of poultry stored at 1-3°C by 8-14 days. Irradiation is of no value as a means of "cleaning up". heavily contaminated food since it would still carry a considerable microbial load, nor does it destroy toxins once they have been produced.

Indeed, in some countries where irradiation is permitted microbiological standards are specified for foods to be treated.

Irradiation of spices to be used for meat products has proved to be an effective way of lowering their microbial content and so increasing the shelf life of the products.

Although the preservation of food by irradiation has been intensively studied for many years its commercial application is still in its infancy. There are many problems involved since the process calls for heavy investment in factory plant and is regarded with some suspicion by consumers. Moreover irradiation does not destroy enzymes so the meat softens during storage.

Since the radiation penetrates into the product it also penetrates packaging so the food can be protected from recontamination by adequate wrapping before irradiation. In general irradiation has some deleterious effect on vitamins but the amount of damage is not considered nutritionally significant.

SIMPLE CANNING

The canning process involves two essential operations: the product must be heated at a sufficiently high temperature and for long enough to make it fully or commercially sterile, and it must be sealed in a hermetic container which will prevent recontamination of the product.

Commercial sterility differs from total sterility in that some organisms may survive the heat treatment but the conditions which prevail in the container during storage do not allow these to grow, produce toxins or spoil the product. However, in regions having a tropical climate, canners strive for total sterility of their canned products. The need to achieve at least commercial sterility determines the minimum heat process to be applied to a product. There is sometimes a temptation for canners to use less than the recommended minimum heat process.

This may result in the product not being commercially sterile. It may then become toxic and poison consumers, or the product may be spoilt and the cans may swell and have to be destroyed. It is essential for people dealing with canning to know what heat processes are required for their products, how these are to be applied, and the nature of the risk they take if less than minimum processes are used. This is why in numerous countries only people having achieved the required level of expertise in canning technology and heat process calculations are authorized to approve heat processes.

Commercial sterility is obtained in meat products which belong to lowacid foods (pH higher than 4.6) if the process applied is severe enough to inactivate the spores. Therefore, meat products are usually processed in steam (or water) under pressure at 116–121°C and sometimes in steam at 140°C. The organisms which are capable of spoiling meat products include those which form heat-resistant spores, thus high-temperature processes are needed to make them commercially sterile.

Usually heat processes for canned products are designed to inactivate large numbers of spores of the organism *Clostridium botulinum.* Although these spores are not as resistant as the spores of some other *Clostridium* and *Bacillus* types, *C. botulinum* is capable of producing lethal toxins, sometimes without swelling the container or obvious alteration of the appearance of the product. Since this organism presents a public health risk, recommended heat treatments must have a large safety margin.

The severity of heat processes for canned meat products is measured in terms of F_0 -values which means that the product received a heat treatment with the same inactivating effect as exposure for one minute at 121°C. For example, one minute at 121°C gives the same amount of inactivation of spores as four minutes at 115°C or 13 minutes at 110°C or 40 minutes at 105°C, so all those processes will have the same F_0 -value. F_0 -value for the majority of canned meat products ranges between one and ten. Larger canned products require higher F_0 -values, even up to 20–25, owing to the longer period needed for heat penetration.

Meat Products Suited for Canning

Meat products made from chopped cured meat that can be canned include corned beef (consisting of large and small pieces of beef, blanched to give a lower water content); corned mutton (similar product made from mutton); beef hash (similar product made from lower-quality meat to which some edible offal is added); minced beef; luncheon meat (similar to emulsiontype sausage) made from beef, pork, or a mixture of beef and pork, or from other kinds of meat; pressed beef (made from large cured high-quality meat pieces; picnic ham (usually weighing 0.5 kg); canned cured ox tongues. Similar products made from pork are: minced pork, chopped pork, lunch pork, pressed pork, pressed ham and many others. Stewed beef, stewed mutton, stewed veal and stewed venison are frequently canned, as well as dishes in jelly such as veal in jelly. Liver sausage, liver paste and some other pastes and frankfurters in brine are also suited for canning.

Containers Suited for Sterilization of Meat Products and Hermetical Closure

Container. The container protects the canned product from spoilage by recontamination with micro-organisms, therefore reliable containers and properly adjusted closing machines are essential to prevent access of spoilage organisms during the cooling operation and during the shelf-life of the product. Before filing, containers have to be rinsed or otherwise cleaned from dust or other impurities. Tin-plate cans are most commonly used, such as the soldered side-seam three-piece can, consisting of can body, can end and lid. The cemented sideseam and welded side-seam cans constitute the majority of cans produced, and remain one of the most reliable and acceptable packages ever conceived. The benefits of using three-piece cans include no size limitations, a wide range of plate thicknesses and tempers for body and ends, abuse resistance, strong end profiles and long shelf-life. Two-piece cans are becoming popular for some applications. Can end and body is drawn in a single piece from flat sheet stock. Only shallow cans (height less than half the diameter) can be fabricated this way. Enamelled cans are more suitable for canned meat products because interior corrosion of cans can be avoided.

Increasingly more attention is being given to aluminium for manufacturing cans and other containers for canned meat products, where special opening features are desired. Important advantages of aluminium cans are that they are lead-free and do not rust. Most often two-piece aluminium cans are used in meat canning. Aluminium for can ends and bodies is, without exception, coated with enamel on both sides. It is widely used in flexible and semi-rigid containers as a protective packaging for a large number of meat products. As a result of extensive development, the use of flexible, laminated pouches and formed aluminium containers for shelf-stable sterilized products is a commercial reality.

The retortable pouch, a thin rectangular package used for heat-sterilized meat products, offers potential improvements in convenience and quality because of its shape and composition. During heat processing, its shape and size allows rapid heat transfer to destroy micro-organisms at the innermost part of the pouch without excessively overheating the product near the pouch wall. The contents are likewise more rapidly cooled at the end of sterilization. A suitable laminate used to make retortable pouches consists of three foils (polyester, aluminium and polypropylene film or resin). These containers are heat-sealed.

Glass containers. Glass is an inert container, although damage to the meat product may result from exposure to light. Glass jars are sealed with caps, twist-off lids and many other systems (crowns, side seal closures, rolled-on closure, screw caps or lug caps) and are used for frankfurters in brine and similar sausages and to some extent also for liver sausage mixtures or sausage mixtures with high-collagen content.

Retortable synthetic casings. Recently a new kind of casing has been used for sausage filling, especially if it is necessary to obtain a longer shelf-life for sausages, sterilized in a retort. This kind of container is sealed with clips made of aluminium.

Smaller size containers are more suitable for meat canning because heat penetration of the meat is by conduction, so if larger containers are used extremely severe heat treatments will be necessary for sterilization. Such severe heat treatment will result in a much lower quality of canned product which will be extremely overcooked. That is why larger pieces, such as hams, shoulders, etc. are pasteurized at temperatures of about 80°C.

Hermetic closure. Tin-plate cans and aluminium cans are closed by machines which form a double seam in three stages: pressing the cover (lid) to the can body, forming the first operation roll seam, and forming the second-operation roll seam.

Good double seams are absolutely essential to prevent losses due to spoilage. The food processor is responsible for the quality of seams produced in the plant and must provide adequately trained personnel to perform seam evaluations and make closing machine adjustments. Making good seams requires careful closing machine maintenance, frequent, regularly scheduled seam evaluations and immediate correction of seam conditions which are outside established tolerances. Hermetic closure of cans under vacuum is highly recommended, particularly for countries with a hot climate.

Autoclave (retort)

Batch retorts are heated with steam or water under pressure. Saturated air-free steam is an excellent heating medium for processing canned meat products. Hot water under pressure has many of the advantages of saturated steam as a heating medium but it must be circulated among containers and continually reheated.

The canning operation must be performed only by fully trained personnel. Operation of batch retorts heated with steam is performed in the following steps:

- load the cans in trays or baskets and close and tighten the retort lid;
- open the drain, vent condensate valves and the high-and low-pressure steam supplies;
- remove the air by a flow of steam during venting;
- close drain and vent valves;

- bring retort to operating temperature and pressure;
- regulate steam supply by temperature control;
- time the process from when the retort reaches the specified temperature;
- open the lid of the retort, remove baskets with canned products and let them dry.

Venting is one of the most important operations in processing canned meat products in steam, because air remaining among cans in an inadequately vented retort forms an insulating layer and will reduce the efficiency of the process.

RAW MATERIALS FOR MEAT PRESERVATION

Product quality and shelf-life of preserved meat and meat products depend on the microbiological and biochemical status of the carcass meat used for processing. In other words, the raw material must be as clean as possible and derived from slaughter animals in good and healthy condition.

Obviously the way of slaughtering animals plays an important role. Many irreversible quality losses, especially with regard to the hygienic quality, originate from improper slaughtering and carcass handling. Slaughter techniques, particularly under rural conditions, cannot and need not be sophisticated, but they must allow carcass meat which complies with basic hygienic requirements to be obtained whether slaughtering at the farm, or in slaughterslabs or slaughterhouses.

Careless slaughter and meat handling result in:

- Improper and insufficient bleeding of the animals, leaving a relatively high amount of blood in the muscle. As a consequence the meat does not reach the necessary degree of acidity and its shelf-life is reduced;
- Improper dehiding of the carcass, leading to heavy contamination of the meat surfaces by frequent contact with the personnel, polluted floors and dirty tools and equipment;
- Improper evisceration through accidental opening of stomachs and tripes, leading to contamination spreading to internal and external surfaces of the carcass;
- Contamination of the meat when carcasses are split on the ground;
- Contamination of the meat during transport, when carcasses or parts of them make contact with unclean clothes, hooks, bars or containers;
- Contamination of the meat stored under unhygienic conditions and because of faulty handling in meat markets, shops and processing plants.

It has to be borne in mind that faulty meat handling, apart from affecting the quality and shelf-life of meat and processed products, may also endanger the health of consumers. Massive contamination will not only enhance meat deterioration caused by food-spoiling bacteria, but may also cause poisoning of consumers by toxin-producing micro-organisms. Toxin-producing micro-organisms find a favourable environment not only in fresh meat stored for a prolonged period at ambient temperatures, but also under certain circumstances in processed meat or even in preserved meat. Appropriate hygienic measures during slaughtering and meat handling are therefore indispensable.

A proper way of improving the slaughter hygiene is to carry out as much of the slaughter operations as possible with the carcass in a hanging position.

Rural Slaughtering on the Farm

Slaughter operations of small animals can easily be performed with the carcass in a hanging position. Problems may arise when slaughtering large animals such as cattle. The bleeding of the animal and cutting off the hind feet is done on the ground and the opening of the skin of the belly and leg region may also have to be done in this position. To complete skinning, evisceration and splitting, simple wooden structures and ropes to gradually hoist up the carcass are strongly recommended.

Remarkable improvements in slaughter hygiene can be achieved by this technique with much less visible contamination (dirt) and a lower degree of invisible contamination. The structure built for slaughtering can also be used for cutting and deboning the carcass.

Slaughterslabs

Slaughterslabs are used in rural communities where the throughput of animals is very small, but where slaughtering in the field should be avoided.

Animals should be suspended from a wooden or metal frame using a gantry hoist to ensure that carcasses are kept off the floor.

The simplest slaughterslab consists of a concrete platform with gantryhoist facilities. The platform should be approximately 50 cm above the ground and sloped to a drain. A roof should be added as protection against adverse weather conditions.

Slaughterslabs should be fenced and roofed and have facilities for hygienic slaughtering and adequate disposal of effluents. If pigs are to be killed a scalding vat should be provided.

Small-Scale Modular Slaughterhouse

The Animal Production and Health Division of the Food and Agriculture Organisation of the United Nations (FAO) has developed a model project for village meat industry in which one of the main components is a smallscale modular slaughterhouse. The design incorporates the use of locally available construction materials and unsophisticated equipment. The slaughterhouse can be built in modules, adding units to the central slaughterhall for operations such as by-product utilisation, meat preservation, processing and butchering. Designs have also been prepared for the construction of a meat market in order to facilitate the integration of production, processing and marketing. It is intended that from the basic nucleus a number of small-scale industries be developed consecutively, *e.g.*, meat preservation by low-cost technologies, traditional tanning, handicrafts based on hides, skin, bone and horns, etc. The objective is to increase employment, in particular for rural women, create market outlets for livestock products and increase the income of small producers.

The project could become a focus for farmer organisation and a centre of technical assistance for producers. The organisation of small producers is indispensable to ensure their participation and for the success and continuity of the project.

The establishment of slaughterhouses of this type in some rural areas because this would increase the availability and quality of meat, improve the utilisation of by-products, avoid rural migration and increase employment. The slaughterhouses may, therefore, constitute a tool for rural development, increasing the income level in the countryside and villages, thus improving the living conditions of the rural communities.

Rural slaughterhouses can also facilitate veterinary control of livestock (ante-mortem inspection) and carcass meat (post-mortem inspection). Slaughtering can be organised to follow a determined daily or weekly time schedule for which compulsory veterinary controls can be arranged. In addition, adequate technical facilities for efficient meat inspection can be provided, *i.e.*, vertical position of the carcasses and special platforms for the personnel. Special hooks or inspection tables can be provided for the inspection of internal organs.

The detailed design for a small-scale modular slaughterhouse. It includes designs, specifications, and schedule of quantities for a slaughterhouse and meat market suitable for small communities.

Provision is made for slaughter of all species: cattle (or buffalo), sheep, goats and pigs, though because of space limitations, concurrent slaughter of different species is not possible. The abattoir capacity will be dependent on the mix of animals being slaughtered. Daily throughputs of approximately 10 large stock (*e.g.*, cattle) or 50 small stock (sheep, goats or pigs) or a combination thereof could be achieved with this design.

The facilities are divided into modules which can be combined as required to suit a particular location.

The following modules are included:

- Lairage
- Slaughter floor
- Chiller
- Tripe room
- Meat cutting and processing room
- Solid waste and blood disposal
- Hides and skin processing
- Effluent disposal
- Electric light and power
- Water supply
- Meat market.

A possible overall abattoir layout based on these modules. This layout shows a typical arrangement for a facility designed to handle beef, small ruminants and pigs. Modifications for a large beef kill and/or the elimination of pig slaughter (*e.g.*, for Muslim communities) are possible. It is evident that when establishing slaughterhouses each country or even separate localities must adopt a solution incorporating special local conditions, locally available materials and manpower, etc. The designs foresee procedures for slaughter of each species as follows:

Cattle

The animal is led into the bleeding area where it is restrained by a tether through the floor ring prior to stunning (using a captive bolt pistol). After stunning the animal is shackled by one leg and hoisted with a rope pulley block. The animal is then stuck and allowed to bleed in this position and the blood collected in a drum for disposal. Once bleeding is complete the head can be removed and the animal lowered on to the cradle for dressing. It is also possible to dress the hanging animal. The feet are then removed, the skin opened up along the breastbone and the hide partially flayed. Leg hooks are attached and the carcass raised to a half-hoist position on the spreader. Flaying can then be completed and the hide removed. The paunch can then be removed to the inspection buggy and the red offal (including lungs if treated as edible) placed on hooks or the inspection table for inspection.

PRESERVATION AND CONSERVATION OF MEAT

The degree of rigour mortis acidification and maturation of the meat largely depend on the condition of the animals and the way they have been treated before slaughtering. It influences the keeping quality of meat and meat products. As such, animals grown for meat purpose should be maintained in good body conditions and should have adequate rest before slaughter. In the muscles of starved animals there is lack of glycogen which means insufficient acidification as a result of which meat does not mature properly.

In pig rearing, specially it is better to give carbohydrates rich food before slaughter so as to increase the quantity of glycogen which favours the formation of lactic acid. Meat of such animals can be suitably preserved since lactic acid improves the keeping quality of meat.

Character of rigour mortis is a good criterion to judge the quality of meat. Rapid development of rigour mortis is an indication that the meat is unsuitable for making preserved preparation as their keeping quality is low.

The muscles of an exhausted animals are poor in oxygen and as such are unable to furnish the required energy for adenosine triphosphoric acid resynthesis. The muscles of exhausted animals do not undergo complete rigour mortis in the absence of hexoses since the breakdown of adenosine phosphoric acid does not occur in a normal way. Adenosine phosphate has a role in the contraction of the muscles. As it is an important criterion, more details have been given about rigour mortis here. It is one of the criteria of keeping quality of meat.

Rigour Mortis

Rigour mortis is setting or firming of muscles. In rigour mortis muscles become hardened, stiffened and rigid. They loose their normal elasticity and extensibility and also become opaque. It is believed that due to the formation of lactic acid rigour mortis occurs and it results in coagulation of myosin. The muscles of the head are affected first and then it spreads backwards over the body. Rigour mortis is a desirable change as it increases the keeping quality of meat, and renders it tender and palatable. Time interval between slaughter and the appearance of rigour, mortis is considerably influenced by muscular activity before death.

In exhausted animals like wild animals hunted to death or animals which are slaughtered by clumpsy methods (that makes them struggle very hard), rigour mortis usually sets in within few minutes of death; the sooner the rigour mortis sets in the quicker the muscles relax again. In such cases, the process of ripening or ageing or conditioning of meat does not proceed in a satisfactory way. In febrile affected animals, rigour mortis does not sets in or manifests very lightly. In animals whose spinal cord and brain are removed immediately after slaughter, rigour mortis passes off more quickly than it would otherwise. Rigour mortis is favourd by high atmospheric temperature and vice versa. The method of slaughter also determines the keeping quality of meat.

Bibliography

Adams, C.E. : *Mammalian Egg Transfer*, Boca Raton, FL, CRC Press, 1982.

Agarwal, Rajesh Kumar: *Meat Hygiene and Food Safety*, Narendra Publishing House, Delhi, 2013.

Archana Satarkar: *Food Science and Nutrition*, ABD Pub, Delhi, 2008.

Arti Sharma: *Fishes : Aid to Collection, Preservation and Identification*, Daya, Delhi, 2006.

Berman, Louis: *Vegetarianism and the Jewish Tradition*. New York: K'tav, 1982.

Brown, L.: *Cruelty to Animals: The Moral Debt*. London: MacMillan, 1988.

Das, Anubrata: *Clean Meat Production with Special Reference to Pork*, Today and Tomorrow's, 2013,

Francis Zimmermann: *The Jungle and the Aroma of Meats : An Ecological Theme in Hindu Medicine*, Motilal Banarsidass, Delhi, 2011.

Haines HR, Sammells : *Adventures in Eating: Anthropological Experiences in Dining from around the World*. University Press of Colorado, 2010.

Hausner, A. : *Preserved Foods and Sweetmeats* , Biotech, Delhi, 2005.

Jhari Sahoo, and Manish Kumar Chatli: *Practical Handbook on Meat Science and Technology*, Daya Pub, Delhi, 2011.

John A.: *Problems With Meat* . Santa Barbara, California: Woodbridge Press, 1979.

Kalechofsky, Roberta: *Haggadah for the Liberated Lamb*. Marblehead, Massachusetts: Micah Publications, 1985.

Kavita Marwaha : *Meat Hygiene*, Gene Tech Books, Delhi, 2007.

Kulkarni, V.V.; A Kalaikannan, S. Suresh Kumar and D. Santhi: *Value Addition and Quality Control of Meat Products : Opportunities and Challenges*, Agrotech Publishing Academy, Delhi, 2014.

Mann N.: *Dietary Lean Red Meat and Human Evolution. Macmillan, Delhi,* 2000.

Montgomery, G. G.: *The Early Placental Mammal Radiation Using Bayesian Phylogenetics* Science, December 2001.

Naveena, B.M.: *Meat Science : A Student Guide*, Satish Serial Publishing House, Delhi, 2013.

Neelam Sachan: *Principles of Meat Technology*, New India Publishing Agency, Delhi, 2011.

Parham, Barbara: *What's Wrong With Eating Meat?* Denver: Ananda Marga Publications, 1979.

Pasewicz M: *Cultural Encyclopedia of Vegetarianism*. Greenwood Press, 2010.

Peterson D.: *Eating Apes*. University of California Press, 2003.

Rifkin, Jeremy: *Beyond Beef - The Rise and Fall of the Cattle Culture*. New York: Dutton, 1992.

Sapontzis SF: *Food for Thought: The Debate over Eating Meat*. Prometheus Books, 2004.

Sharma, Davinder Kumar: *Preservation of Meat and Poultry Products*, Asia Pacific Business Press, 2005

Shoshan, A.: *Man and Animal* (Hebrew). Jerusalem: Shoshanim, 1963.

Vijay J. Jadhav: *Handbook of Quality Control of Dairy and Meat Products*, Biotech Books, 2012.

William D. Weaver : *Commercial Chicken Meat and Egg Production*, Springer, Delhi, 2007.

Index

N

P

R

S

V

W

❑❑❑